Introduction to
Infrared System Design

TUTORIAL TEXTS SERIES

Introduction to
Infrared System Design

William L. Wolfe
Professor Emeritus, Optical Sciences Center, University of Arizona

Tutorial Texts in Optical Engineering
Volume TT24

Donald C. O'Shea, Series Editor
Georgia Institute of Technology

SPIE OPTICAL ENGINEERING PRESS

A Publication of SPIE—The International Society for Optical Engineering
Bellingham, Washington USA

Library of Congress Cataloging-in-Publication Data

Wolfe, William L.
     Introduction to infrared system design / William L. Wolfe.
      p. cm. -- (Tutorial texts in optical engineering ; TT 24)
     Includes bibliographical references and index.
     ISBN 0-8194-2106-5 (softcover)
     1. Infrared equipment--Design and construction. I. Title.
  II. Series.
TA1570. W65 1996
621. 36'2--dc20   96-5444
                                    CIP

Published by
SPIE—The International Society for Optical Engineering
PO Box 10
Bellingham, Washington 98227-0010
Phone: 360/676-3290 Fax: 360/647-1445
E-mail: spie@spie.org
WWW: http://www.spie.org/

Printed in the United States of America.

Second printing.

SERIES INTRODUCTION

The Tutorial Texts series was begun in response to requests for copies of SPIE short course notes by those who were not able to attend a course. By policy the notes are the property of the instructors and are not available for sale. Since short course notes are intended only to guide the discussion, supplement the presentation, and relieve the lecturer of generating complicated graphics on the spot, they cannot substitute for a text. As one who has evaluated many sets of course notes for possible use in this series, I have found that material unsupported by the lecture is not very useful. The notes provide more frustration than illumination.

What the Tutorial Texts series does is to fill in the gaps, establish the continuity, and clarify the arguments that can only be glimpsed in the notes. When topics are evaluated for this series, the paramount concern in determining whether to proceed with the project is whether it effectively addresses the basic concepts of the topic. Each manuscript is reviewed at the initial state when the material is in the form of notes and then later at the final draft. Always, the text is evaluated to ensure that it presents sufficient theory to build a basic understanding and then uses this understanding to give the reader a practical working knowledge of the topic. References are included as an essential part of each text for the reader requiring more in-depth study.

One advantage of the Tutorial Texts series is our ability to cover new fields as they are developing. In fields such as sensor fusion, morphological image processing, and digital compression techniques, the textbooks on these topics were limited or unavailable. Since 1989 the Tutorial Texts have provided an introduction to those seeking to understand these and other equally exciting technologies. We have expanded the series beyond topics covered by the short course program to encompass contributions from experts in their field who can write with authority and clarity at an introductory level. The emphasis is always on the tutorial nature of the text. It is my hope that over the next five years there will be as many additional titles with the quality and breadth of the first five years.

Donald C. O'Shea
Georgia Institute of Technology

CONTENTS

PREFACE

The notes that were the basis for this tutorial text have been compiled over a period of about thirty years. During that time there have been many changes in the field of infrared technology. For the most part these changes have been incorporated in the basics, for basics are basic. I have had the privilege of investigating many different problems, either as an industrialist, in my former incarnation at the Honeywell Radiation Center (now Loral), as a teacher, and as a consultant. In the latter role I have worked for both the government as a critic and design reviewer as well as a creator.

Each of the design examples originated as a real problem. I have altered them to protect the innocent and to make the problems more interesting. This was done chiefly in the *Peacekeeper* and ICBM detector problems to make the choices among the options somewhat less obvious.

I have concentrated on the optics and detector aspects of infrared system design. Although the mechanical design of the structures is equally important, the techniques are those normally taught to the mechanical engineer and are not peculiar to infrared systems. A similar situation is true with respect to the electronics of the system.

I am indebted to those who first challenged me with these. I am also indebted to my past three employers—The University of Michigan, Honeywell, and The University of Arizona—for keeping me employed all those years. (Sometimes you can fool all of the people all of the time!)

I am also indebted to SPIE for the fine job they did in the publication. This is due mainly to Mary Kalbach Horan, the eloquent editor, Don O'Shea, the persistent and perspicacious Series editor, and George Zissis, the careful and consummate critic—and friend.

I wish to dedicate this book to my patient wife of over forty years, Mary Lou, who frequently asked, "And what are you doing in front of that computer today?"

My long-time friend Stan Ballard said, "It is nice to have written a book." He was right.

William L. Wolfe
Tucson, Arizona
1996

SYMBOLS & NOTATION

α	absorptivity	F	optical speed
α	angle	g	recombination factor
β	absorption coefficient	G	optical gain
β	resolution angle	GCF	geometric configuration factor
γ	angle		
Δf	bandwidth	h	Planck's constant
ϵ	emissivity	i	image distance
η	quantum efficiency	k	radian frequency
η	efficiency	K	proportionality constant
θ	pixel angle	L	radiance
Θ	field angle	m	detector number
λ	wavelength	M	exitance (emittance)
ν	frequency	M	modulation
π	3.14159...	MRTD	minimum resolvable temperature difference
ρ	reflectivity		
ρ	distance	n	refractive index
σ	wave number	N	number
σ	Stephan Boltzmann constant	NA	numerical aperture
τ	transmissivity	NEP	noise equivalent power
τ	time constant	NETD	noise equivalent temperature difference
ϕ	pixel angle		
ω	radian frequency	o	object distance
Ω	solid angle	q	electronic charge
Ω'	projected solid angle	$\mathfrak{R}$	responsivity
A	area	R	range, distance
B	bandwidth	$R(\lambda)$	general radiometric quantity
c	speed of light	SNR	signal-to-noise ratio
C	contrast	t	time
c_1	first radiation constant	T	temperature
c_2	second radiation constant	TDI	time delay and integration
D	detectivity	Th	throughput
D^*	specific detectivity	U	energy
E	incidance (irradiance)	V	voltage
f	focal length	x	dimensionless frequency
f	frequency	y	displacement
F	Focal point	Z	throughput

SUPERSCRIPTS AND SUBSCRIPTS

$\Delta\lambda$	spectral band
1,2	first, second...
a	atmospheric
AA	astigmatism
b	bidirectional
BB	blackbody
CA	coma
d	detector
g	gap
h	horizontal
h	hemispherical
i	incident
m	detector number
m	maximum
max	maximum
ms	mean square
n	noise
o	optics
os	optics at source
q	photon
r	reflected
s	source, signal
SA	spherical
so	source at optics
t	transmitted
v	vertical, visible

Introduction to
Infrared System Design

1 INTRODUCTION

The infrared spectrum has provided a rich, available technology for the accomplishment of many important and practical tasks. This tutorial text provides a brief introduction to the language, processes, and some of the instrument design techniques that are available to the engineer today. It arose from lectures given at about ten different tutorial sessions at meetings of SPIE, the International Society for Optical Engineering. The material was adapted from a full, one-semester, three-hour course, given at The University of Arizona for twenty five years.

The infrared spectrum is part of the electromagnetic spectrum, with wavelengths ranging from about 1 μm to about 1000 μm (1 mm). It lies just above the visible spectrum that spans the region from 0.3 μm to 0.8 μm (300 nm to 800 nm) and below the millimeter region. There are various names for the various parts of the infrared spectrum, and they are used differently by different people. Most people would consider the far infrared as ranging from about 25 μm to 1000 μm. It is used chiefly by astronomers and solid state physicists. That part of the region will not be covered in this text. The remaining part of the spectrum, ranging from 1 μm to 25 μm, is divided into the short-wave infrared (SWIR), from 1 to 3 μm, the midwave infrared (MWIR), from 3 to 5 μm, and the long-wave infrared (LWIR), from 8 to 12 μm. The remaining region, from 12 to 25 μm, is not used a great deal, but has been called the very long wave infrared (VLWIR). These regions are defined by the transmission of the atmosphere and the spectral sensitivity of various detectors. The last-named region is used only above the atmosphere.

The general plan of this text is to discuss first some of the existing applications of infrared as a sort of appetizer and incentive to learn the techniques discussed in the later chapters. Then a review of optical fundamentals, the requisite radiometry, and of detector types and properties is given. The main body begins with a rather mathematical treatment of the equations of sensitivity. These are done in two different ways. The first is in terms of the summary figure of merit, specific detectivity, that has been used and misused for almost a half century. It is useful in many applications; it does have limitations that must be recognized. The second method is the counting of photoelectrons by charge-collection devices. Idealized equations for signal-to-noise ratio and noise-equivalent temperature difference are developed and discussed. A review is then given of the nature of the infrared scene. The infrared system designer must keep in mind that the world around him is rich in infrared radiation, and the infrared characteristics of all these objects can be vastly different than their visible counterparts. Snow is black in the infrared! This information can then be incorporated appropriately in the sensitivity equations. The radiation from the scene is transmitted, absorbed, and scattered by the atmosphere as it travels to the sensor. Techniques for making these calculations are then discussed. The information and noise bandwidths are also important in sensitivity calculations. They are covered next, along with a variety of scan patterns and techniques, but not scanner realizations.

The following three chapters discuss the properties of optical materials for windows, lenses, and mirrors, some representative telescope systems, and optical-

1

mechanical scanning techniques. The final chapter pulls these concepts together by discussing the approaches to several different applications, a real-time imager, a strip mapper, and point-source surveillance systems. These examples show how to approach the problems, but they must be viewed as incomplete solutions. Appendixes A and B provide some useful programs in QuickBASIC that are readily adaptable to Visual BASIC and to other scientific programming languages. The volume closes with a bibliography, a collection of most of the useful texts with comments about them all.

2 APPLICATIONS OVERVIEW

The infrared part of the spectrum has been used for many different scientific, military, medical, forensic, civilian, industrial, practical, astronomical, and microscopic applications. One good place to read more about all this is Hudson.[1] Another is the September 1959 issue of the Proceedings of the Institute of Radio Engineers (Proc. IRE), now Proceedings of the IEEE. We will discuss just enough of these to give the flavor of the diversity. Astronomical applications have been in the news in large measure because of the Infrared Astronomical Satellite (IRAS) and Shuttle Infrared Telescope Facility (SIRTF). Because the infrared part of the spectrum gives much thermal information and other clues that the visible does not, it is an important part of the astronomical observational arsenal. Long wavelengths are important, as is rising above the absorptions of the atmosphere. Our weather system is gradually being understood better as a result of many infrared measurements of temperature profiles and cloud distributions. The railroads have benefited from hot-box detectors. These devices, as simple as they were, can be placed by the side of the tracks to look for hot spots that are indicative of journal boxes that have lost their lubricants. Houses can now have infrared warning and intrusion detectors. They can also be analyzed for the efficacy of their insulation. Although they have been investigated, aircraft collision warning systems using infrared have not been introduced.

Power-line relay poles have been investigated in several areas. The increase in temperature associated with the higher resistance of poor contacts (probably that have weathered) is easily detected. The problem is covering enough area. Fire fighting in remote areas has been aided by infrared reconnaissance. The fire, under a canopy of smoke, can be located quite well by infrared techniques.

Thermography, usually understood in the medical sense, has contributed to breast-cancer detection, to burn problems, to circulatory analyses, and even to the diagnosis of malingerers. This can be applied sometimes when back pain is claimed and there is no sign of the telltale temperature increase. Almost all pathology is closely associated with an increase in temperature: the thumb we hit with a hammer surely gets hot. In the military, infrared has been used for night vision, missile homing devices, tail-warning systems, ballistic missile detection, other kinds of intrusion detection, and many other applications. The dramatic example of the missile going down the pipe of the Iraqi bunker is a spectacular example of this.

The efficiency of my automobile engine is now monitored about once a year. They stick an infrared sensor up its . . . tailpipe, and they sense the amount of carbon monoxide and carbon dioxide, as well as other hydrocarbons and uncombusted products, as an indication of the completeness of combustion. Some breathalysers are infrared spectrometers in disguise. And it has even been proposed that police look deep into our eyes with infrared to identify us (because it is said that the back of the eye is as unique as a fingerprint). Some automobile engines are made with the assistance of infrared

---

[1] R. D. Hudson, *Infrared System Engineering*, Wiley, 1969.

devices that can monitor the weldments and castings for both uniformity and for absolute temperature.

The now-familiar automatic door openers and intrusion warning systems are another clever example. They use a plastic Fresnel lens and one or two thermal detectors to sense anything that is warm and moving. Infrared imagers have been used in the Marana, Arizona, copper smelter to assess the efficiency of the electrodes that are used in the purification process (where heat is again an indicator of poorer conduction). This process has improved the efficiency by about 4%, but this is a gigawatt operation! A gigawatt here, a gigawatt there soon adds up to real savings!

3 REVIEW OF GEOMETRIC OPTICS

Certain elementary concepts in optics are necessary to appreciate infrared system design both for radiometric and resolution calculations. Only those concepts necessary are discussed here, and they are not derived. Further reading is available in several texts.[1]

3.1 Rays, Beams, and Pencils

A ray is defined as the normal to wavefront, and describes the path of the light in a geometric fashion. A beam is a collection of rays. Beams can be collimated, in which case all the rays are parallel. They can be divergent or convergent, in which case they emanate or converge to a point with a geometry that looks like the sharpened end of a pencil. These are called conical beams.

3.2 The Laws of Reflection and Refraction

The refractive index of a medium is defined as the ratio of the velocity of light *in vacuo* to that in the medium. All media have a refractive index greater than one. The refractive index of a mirror is −1 by convention. The light goes at the same speed, but in the opposite direction.

When light is incident from a medium of a particular refractive index onto a medium that has a different refractive index, it is both reflected and refracted at the surface between them. This is illustrated in Fig. 3-1. The direction of the (specularly) reflected beam is determined by θ_i and θ_r, the angles of incidence and reflection. These are both measured with respect to the surface normal, as shown. The angle of reflection is equal to the angle of incidence.

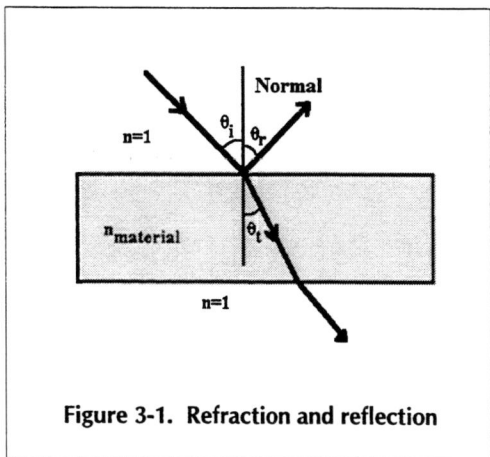

Figure 3-1. Refraction and reflection

The direction of the refracted beam, the one that is in the medium and is deviated to some extent from the incident beam, is determined by Snell's law: the sine of the angle of refraction times the refractive index of the medium is equal to the sine of the angle of incidence times the refractive index of the incident medium.

---

[1]F. A. Jenkins and H. E. White, *Fundamentals of Optics*, McGraw-Hill, 1957; J. Strong, *Concepts of Classical Optics*, Freeman, 1958; A. Hardy and F. Perrin, *The Principles of Optics*, McGraw-Hill, 1932.

$$n_{air} \sin \theta_i = n_{material} \sin \theta_t \, , \tag{3.1}$$

where θ_i is the angle of incidence, as shown in the figure, θ_r is the angle of reflection and θ_t is the angle of refraction. The refractive index of air is taken as 1, and that of the material must be measured or specified.

3.3 Gaussian Optics

Gaussian optics is a synonym for perfect optical imaging in which there is one-to-one, linear mapping of an object to an image. Figure 3-2 illustrates this relatively well known geometry and also defines the front and rear focal points F and F' the optical axis, and the object and image planes. Figure 3-2 shows a thin lens, one that has refracting properties as if it had curved surfaces, but it has no thickness at all. The optical axis (shown by a dashed line) passes through the center of the lens and both focal points. The back focal point is the point from which rays diverge and leave the lens parallel to the optical axis. The front focal point is analogous to this; a ray that passes through it (and goes through the lens) will exit the lens parallel to the optical axis. The focal distance or focal length f is the distance from the lens to the focal point. An object is illustrated by the vertical arrow y on the left. Its image y' is the inverted arrow on the right. The image of the object may be determined by tracing a ray parallel to the optical axis from the object that goes through the rear focal point, and a ray that goes through the center of the lens and is undeviated. It can also be determined by the thin lens equation

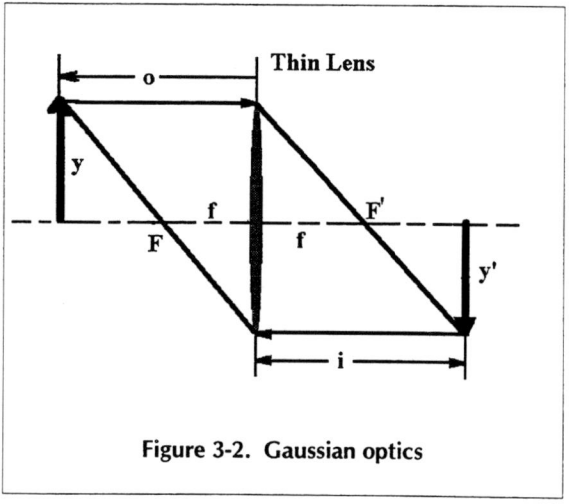

Figure 3-2. Gaussian optics

$$\frac{1}{f} = \frac{1}{o} + \frac{1}{i} \, , \tag{3.2}$$

where f is the focal length, o is the object distance, and i is the image distance. Other authors use other symbols, *many* other symbols.

The first step in optical design starts with this thin-lens equation. The second step uses the lens makers' equation,

$$\frac{1}{f} = (n - 1)\left(\frac{1}{R_1} + \frac{1}{R_2} \right) , \tag{3.3}$$

where n is the refractive index of the lens, while R_1 and R_2 represent the radii of curvature of the first and second surfaces. The usual sign convention for these radii is that if the center of curvature is to the right of the surface, the radius is positive, and when it is to the left, it is negative. That convention will be used here.

3.4 Paraxial Optics

So-called paraxial optics uses the assumption that the rays are parallel (or almost parallel) to the optical axis, that is, the rays make small angles with the optical axis. This is a linearization in which sines and tangents are approximated by their angles (in radians). The resulting equations are simple, but even these are more complicated than necessary for present purposes; the thin-lens equations and the third-order approximations for aberrations will be enough.

3.5 Stops and Pupils

Stops and pupils control the amount of light that is accepted by an optical system and the size of the field of view. There are two types of optical stops: aperture stops and field stops. These are illustrated in Fig. 3-3. The aperture stop determines the size (diameter or area) of the beam of light that is accepted by the optical system. It is often the first element of the optical train, but it does not have to be. A field stop determines the angular extent over which the beams can be accepted. Pupils are images of stops; the entrance pupil is the image of the aperture stop formed by all elements in front of it. If the aperture stop is the first element, it is also the entrance pupil. The exit pupil is the image of the aperture stop formed by all elements behind it. The entrance and exit windows are the corresponding images of the field stop. Note that a stop must be a physical element like the edge of a lens, a detector, or an iris, but the pupils need not be physical entities.

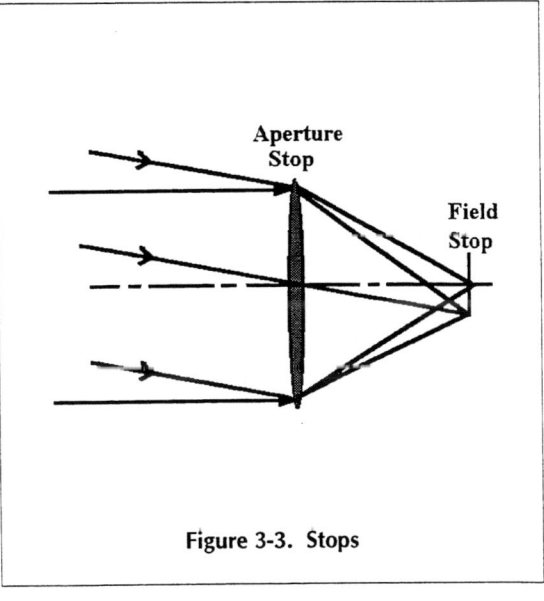

Figure 3-3. Stops

3.6 Optical Speed

The speed of an optical system is often expressed in terms of the focal ratio or the numerical aperture. The focal ratio, also called the F number and signified in this text by F, rather than the often cited $F\#$, but also signified by FN, $f/\#$, $f\#$, $F\#$, and more, is the ratio of the focal length to the diameter of entrance pupil. It is a measure of the convergence of light onto the image plane. The numerical aperture, almost universally denoted by NA, is defined as the refractive index times the angle of the convergence. It is another measure of the convergence; the two are related for infinitely distant objects by

$$NA = \frac{1}{2F} .$$

$$(3.4)$$

3.7 Aberrations

Aberrations are deviations of real rays from the idealized, paraxial assumptions made above. Third-order aberrations are those that are calculated by adding the cubic (third-order) term to the angle in the series for sines and tangents. Fifth-order aberrations include the next term, etc. Only third-order aberrations will be considered in this text.

3.7.1 Spherical Aberration, Spherical

A spherical surface is not the right surface to bring parallel light to a focus on axis. The general nature of spherical aberration is shown in Fig. 3-4. Spherical aberration is independent of field angle. It is given, for an element which has its aperture at the element, as

$$\beta_{SA} = \frac{1}{128F^3} \left[\frac{n(4n-1)}{(n+2)(n-1)^3} \right] ,$$

$$(3.5)$$

where β is the angular diameter of the aberration, F is the F number, and n is the refractive index of the lens. If the optical element is a mirror, the value in the square brackets is one.

3.7.2 Comatic Aberration, Coma

Comatic aberration occurs only for field locations that are not on axis. It occurs because the magnification of the lens (mirror) is not constant for all field angles. A similar expression for comatic aberration is

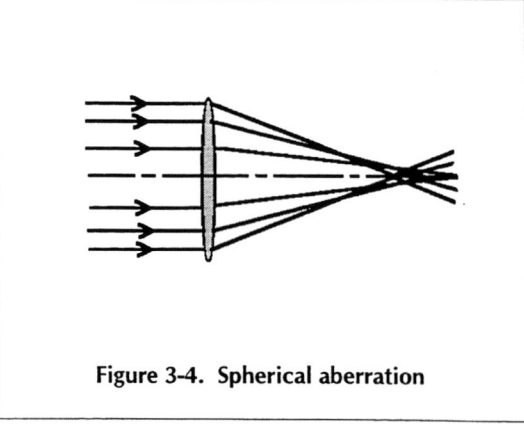

Figure 3-4. Spherical aberration

$$\beta_{CA} = \frac{\Theta}{16F^2}\left[\frac{1}{n+2}\right],$$ (3.6)

where the variables are the same, but Θ is the maximum angle a ray in the field of view makes with the axis. Again, the bracket is unity for a mirror.

3.7.3 Astigmatic Aberration, Astigmatism

Astigmatism is obvious further off axis. It is a result of the additional asymmetry between the horizontal (tangential) and vertical (sagittal) performance of the lens. Its equation is

$$\beta_{AA} = \frac{\Theta^2}{2F}.$$ (3.7)

Both lenses and mirrors behave the same.

3.7.4 Curvature of Field

In general, the image of the full field of view is curved, as shown in Fig. 3-5. This curvature is caused by the curvature of the components. There are design maneuvers to reduce this to a minimum. One is to attempt to make all the curvatures of the surfaces, with due account of signs, add to zero. This doesn't insure that the image will not be curved, but it helps. Another approach is to add a field-flattening lens in the system. It is deleterious to have the field curved, whether the detector is a piece of film or a detector array. In some infrared systems, very large ones, the curvature is approximated by segments of arrays of detectors.

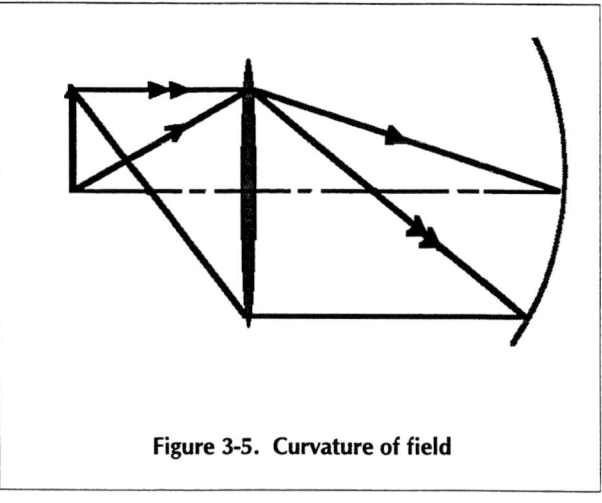

Figure 3-5. Curvature of field

3.7.5 Distortion

Imaging is not linear, and in general, the separation of equally spaced points in the object is not maintained; the separation either increases or decreases with increasing field angle. This can cause mensuration problems in certain types of seekers and sensors.

The distortion is generally called either barrel or pincushion, depending on the shape of the image of a true square. Distortion can be an especially nasty aberration if an imaging system is used for tracking targets, because the mensuration algorithms become tainted by the distortion. Computer correction is possible, but it can get computation intensive.

3.8 Combinations of Aberrations

The aberration expressions given above are each for a single optical element, lens, or mirror. If two or three (or more) are used, how does one combine their individual aberrations? Options are simple addition or root sum square "addition." Neither is correct, because in the hands of a good optical designer, a second element can actually correct for the aberrations caused by the first. In my recommendations for optical scheming, I suggest that the calculations be made as if the entire system were a single element. If the results are about right, even a little pessimistic, continue with the design. A good optical designer will help you out. A nice treatment of two-mirror systems has been given by Gascoigne.[2] It is recommended as further reading for those interested in this subject.

---

[2] G. C. B. Gascoigne, "Recent advances in astronomical optics," *Applied Optics,* **12,** 1419 (1973).

4 REVIEW OF RADIOMETRY

Radiometric calculations for the generation and transfer of radiation are reviewed in this chapter. Every infrared system senses the power that is incident upon the detector. This power is generated by a remote source and transmitted to the optical system, which in turn focuses it on the detector. All these aspects of signal generation come under the general heading of radiometry. Therefore, this chapter reviews the pertinent topics.

These topics include the concepts of solid angles, projected solid angles, throughput, elements of radiometric transfer, definitions of radiometric quantities for both fields and materials, and some properties of blackbodies. It is necessary to define several radiometric terms to describe how radiation transfers from sources to receivers. There has been much ado about the words and symbols associated with these concepts, and we will spend no more time than is absolutely necessary with these excursions into semantics. There are four different geometric terms related to power transfer—in addition to power. They all have some relationship to a solid angle, so it is important to obtain a clear definition of solid angle. This may be done by reviewing first the definition of an ordinary or linear angle.

4.1 Angle and Solid Angle

An angle can be defined as the length of arc of a circle divided by the radius of the circle, as shown in Fig. 4-1. It can be measured in degrees, and by definition there are 360 degrees in a circle. This is an arbitrary measure, first promulgated by the ancient Babylonians. The natural measure of an angle is as a fraction of the circumference of the circle. Such a measure is called a radian. The radian is the ratio of the arc of the circle to the radius of that circle. The circumference of the circle is 2π times the radius, so there are 2π radians in a circle. But a line can also subtend an angle, and it may not be tangent to the circle or perpendicular to the radius. In such a case, the perpendicular component must be taken to assess the angular subtense of the line at a given point. The figure also shows a line that

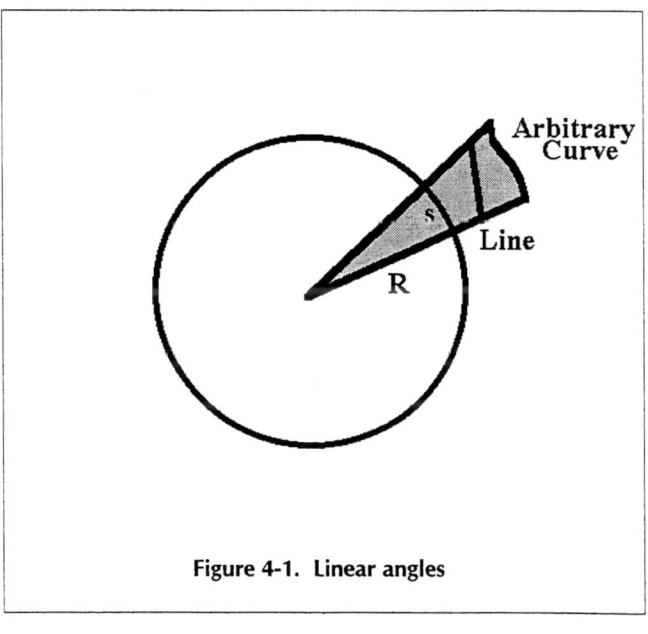

Figure 4-1. Linear angles

subtends the same angle as the arc, mentioned above, and an arbitrary curve that does the same. Any arbitrary line must be projected onto the circle to obtain the proper measure of the angle.

The solid angle is defined as an element of the area of a sphere divided by the square of the radius of the sphere. It represents a fraction of the entire surface of the sphere. Figure 4-2 is a representation of a solid angle. It can be measured in square degrees, but the natural measure is square radians which are called steradians, meaning solid radians. The area of the surface of the sphere is 4π times the square of the radius, so there are 4π steradians in a sphere. I do not believe I have ever heard anyone state

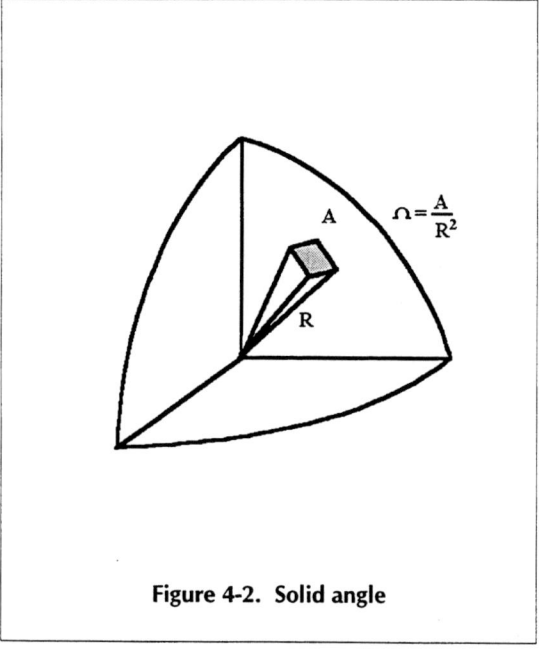

Figure 4-2. Solid angle

how many square degrees there are in a full sphere, although sometimes (relatively small) solid angles are measured in square degrees. A hemisphere, logically enough, contains half as many radians as a sphere, that is, 2π. Other areas can also subtend solid angles, but like the lines they must be measured as *projected* perpendicular to the line from the area to the point of evaluation.

4.2 Spectral Terms

The radiometric terms may be separated into geometric terms and spectral terms. A radiant power may be the power in a spectral band, or it may be the power per certain spectral interval. The spectral term is usually denoted by wavelength λ in the field of infrared system design, although it is also specified as the wave number, denoted by σ or $\bar{v}$. The wavelength is usually specified in micrometers, μm, one millionth of a meter; the wave number as reciprocal centimeters. Rarely is the radian wave number k (=$2\pi \sigma$) or the frequency v(Hz), f(Hz), or x (dimensionless) used. In this text, λ and σ will be used.

4.3 Radiometric Terms

The first and most fundamental of radiometric terms is *radiance*. It is the radiant power per unit projected area and per unit solid angle. This may also be considered the power that is distributed with respect to area and solid angle. The integration of radiance over solid angle yields the power per unit area or flux density. If the power leaves the surface, it is usually called exitance (or emittance); if it is incident upon the surface it is called incidance (or irradiance). The integral of radiance over area gives the power per unit solid angle, which is usually called intensity.

Radiance has the units of watts per unit area per unit solid angle, usually in the units $Wcm^{-2}sr^{-1}$, and is designated by the symbol L. Similarly, radiant exitance (emittance) and incidance (irradiance) have the units of power per unit area, Wcm^{-2}, and are designated as M and E respectively. Radiant intensity has the units of Wsr^{-1}, and it is designated by I. Until about thirty years ago, the preferred symbols were N, W, H, and J, instead of (correspondingly) L, M, E, and I. In this scheme, power is indicated by Φ for flux (Φlux?). I will use this set of symbols. Some reports still have the older symbols. Some of us think the older symbols were better, but that is the way it is. Some authors (and others) have proposed alternate words for these. The terms defined above are listed here with their alternates: radiance (sterance), intensity (pointance), exitance (emittance, areance), and irradiance (incidance and areance). It is not important why, but these are the terms that will sometimes be found in the radiometric literature.

Table 4-1 summarizes these quantities. The first column is the general quantity. The second indicates the quantity name and units for energetic radiation. The second does the same for photonic radiation (photons per second, photons per second per area, etc.). The third does the same for visible radiation. The last column is a set of terms suggested by Fred Nicodemus, but they have never received widespread acceptance. They were used by Wyatt in his texts on radiometry (see the references in Appendix A).

Table 4-1. Radiometric Terms and Nomenclature

Quantity	Energy U	Photons N	Visible U_v	Nicodemus
Volume density	u	n		
Flux Φ	Power Φ	Rate N	Lumens Φ_v	Flux
Flux density				Areance
Incidance E	Irradiance E_u	Phincidance E_q	Illuminance E_v	Incidance
Exitance M	Emittance M_u	Phittance M_q	Luminous Emittance E_v	Exitance
Sterance	Radiance L_u	Photance L_q	Luminance L_v	Sterance

Each of these represents a certain geometry. In addition, the spectral characteristics may be designated for any of them by subscripting with a λ or other spectral variable to indicate that the quantity is spectrally distributed. The units in that case would include also a "per whatever the spectral variable is." For instance, the spectral radiant intensity might have units of watts per steradian per micrometer for a wavelength distribution, or watts per steradian per wave number for a frequency distribution. A spectral band can be indicated by, for instance, $d\lambda$, and the visible weighted quantity may be indicated by a subscript v. Photon rates may be designated by a subscript q, although some use a p. They can also be used in combination; for example, $I_{q\lambda}$ is the spectral photon intensity with units of photons per second per steradian per micrometer.

The properties of materials have special names as well. These include endings of *ion, ance,* and *ivity*, each with a special meaning. The first means a process; the second, a property of a sample; the third, a generic property of this material in which the surface structure, degree of polish, etc., do not contribute to this inherent property. It seems impossible to measure an *ivity*, for no sample is prepared well enough. This scheme, however, does require that emittance be the radiation efficiency of a particular sample, dimensionless, and not the emitted flux density from a sample. Therefore other terms were invented, like *exitance* for the flux density. In addition, it was recognized by Jones that most of the system of radiometric nomenclature was generic to any quantity. One can have a quantity, like energy or photons, and its flux (time rate), its flux density, which he called *exitance* and *incidance*, its intensity (still intensity), and its radiance, which he called *sterance*. Thus came a new set of names—and still more from Nicodemus, who introduced *pointance* for intensity and areance to include both exitance and incidance. These nuances will not be pursued or followed in this text.

4.4 Radiometric Transfer

The fundamental equation of radiation transfer (in a lossless medium with unity refractive index) states that a differential element of power is given by the radiance times the product of two projected areas, divided by the square of the distance between them:

$$d\Phi = L\frac{dA_1 \cos\theta_1 \, dA_2 \cos\theta_2}{\rho^2} .$$

(4.1)

This is illustrated in Fig. 4-3 and represented by Eq. (4.1). For many situations this can be simplified to the areas of sources, lenses, and the like, the cosines can be dropped, and a set of useful relationships arises. These are notably for the point (subresolution) source and the extended source. This fundamental equation of radiation transfer can be rewritten in terms of the projected area and the solid angle or the other projected area and the other solid angle.

$$d\Phi = LdA_1 \cos\theta_1 \, d\Omega_2 = Ld\Omega_1 \, dA_2 \cos\theta_2 .$$

(4.2)

The equation is completely symmetrical. It can be used to calculate the power flow in either direction or the net power flow. Equation (4.2) is partly due to the definition of radiance, which is the flow per unit area and per unit solid angle. Whenever there is a question about which solid angle to use, return to the fundamental definition, using both of the projected areas. A further definition involves the projected solid angle. The projected solid angle is usually designated by Ω', but sometimes by Ω_p. The fundamental equation becomes

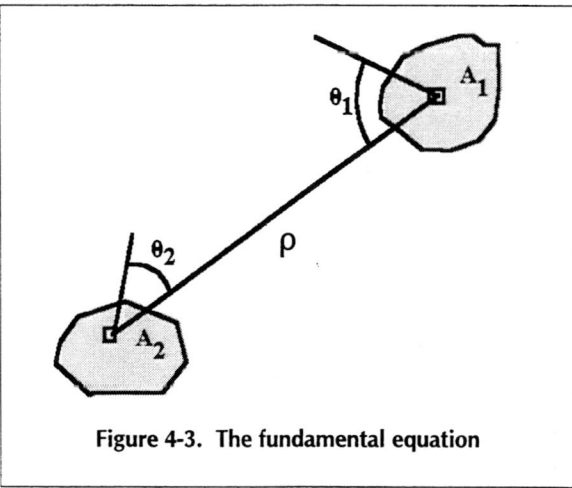

Figure 4-3. The fundamental equation

$$d\Phi = LdA_1\, d\Omega'_2 = LdA_2\, d\Omega'_1 .$$
(4.3)

Some authors use a lowercase omega, ω, to indicate solid angle and an uppercase omega, Ω, to indicate projected solid angle. The projected solid angle has no physical significance. It is simply the quantity one multiplies area and radiance by to get power, but it is useful and used. The fundamental equation can also be written as

$$d\Phi = LdZ = LdTh = LdA\Omega .$$
(4.4)

Here the form is that of the radiance times a quantity that includes all the other terms. Together they are called the throughput, optical extent, etendué, the A-Ω product (and certain four letter words). It is a property of the geometry alone, while the radiance is a property of the field alone. A nice separation is thereby obtained. It is the significant quantity involved with the radiometry of extended objects. It is a constant for a given geometry and for a given optical system.

For isotropic radiators, those that have the same radiance in all directions, the radiance is just the radiant exitance [emittance] divided by π, and another form is possible. The power transferred is the radiant exitance times the geometric configuration factor (GCF), or it is the radiance times the GCF divided by π.

$$P = LZ = \frac{M}{\pi}Z = M\,\text{GCF} .$$
(4.5)

The GCF is also known as the angle factor, and it is compiled for many geometries in mechanical engineering books related to radiative transfer.[1] These have become almost obsolete, partly as a result of computers and partly because they are limited by the Lambertian assumption.

4.4.1 Lambertian Emitters (and Reflectors)

The fundamental equation of transfer can be applied to a Lambertian emitter to obtain the relationship between radiance and radiant exitance (emittance) for such a source. This radiator emits uniformly into an overlying hemisphere. A Lambertian radiator or even reflector is one that emits or reflects a radiance that is independent of angle; the radiance is isotropic. This property is only the figment of a theoretician's imagination; no material is truly isotropic in either emission or reflection. One assumes that there is a differential area, dA_1, of emitter that has a radiance independent of angle. The fundamental equation of radiative transfer is applied, as shown in Fig. 4-4.

$$\Phi = L\frac{dA_1 \cos\theta_1 \, dA_2 \cos\theta_2}{\rho} = L\frac{dA \cos\theta R \, d\phi R \sin\theta \, d\theta}{R^2} \tag{4.6}$$

In this case, we can drop the subscripts because the second area is a differential area on the surface of the sphere of value $Rd\phi R\sin\theta d\theta$. The surface is perpendicular to the radius, so that $\cos\theta_2$ is just equal to 1. The distance between elements is a constant, R, rather than a variable ρ. The differential area and the radius are both independent of the angles of integration, and the radius R cancels out of the numerator and denominator. The result is an integration over ϕ, which is 2π, and an integration over θ, which gives a solid angle of $\pi\sin^2\theta/2$, which for a hemisphere is just π.

$$M = \frac{P}{dA} = L\int_0^{2\pi}\int_0^{\pi/2}\sin\theta\cos\theta \, d\theta \, d\phi$$
$$= 2\pi L\int_0^{\pi/2}\sin\theta\cos\theta \, d\theta \; . \tag{4.7}$$

$$M = L\pi[\sin^2\theta]_0^{\pi/2} = \pi L \; . \tag{4.8}$$

Thus the radiant exitance is just π times the radiance. Since it is uniform over the hemisphere, i.e., isotropic, it may be inferred that the value of the projected solid angle of a hemisphere is π, while the value of the solid angle is 2π. The difference is a factor of $1/2$, the average value of the cosine over the half cycle. This may also be interpreted

---

[1]R. Siegel and J. Howell, *Thermal Radiation Transfer*, McGraw-Hill, 1971, and references therein.

as the average value of the projected area of the source over the hemisphere's angular excursion.

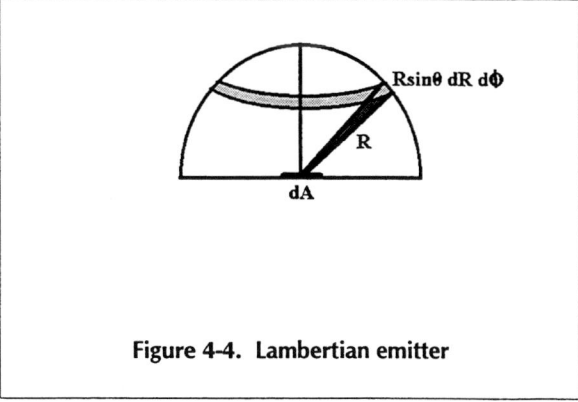

Figure 4-4. Lambertian emitter

The same analysis applies to the reflection of light by a Lambertian (isotropic) reflector. The incidance (irradiance) on the sample will be converted to reflected radiance that is reflected uniformly in all directions with a radiance given by the reflectivity times the incidance (irradiance) divided by π.

Although it is not proven here, the solid angle is determined only by the periphery of the area—in this case, the sphere. This is similar to a linear angle, where the angle depends on the ends of the line and not at all on the shape of the line. Various techniques have been used to calculate all sorts of solid angles and projected solid angles using these ideas, but they cannot be presented here, nor are they necessary to obtain further results in infrared system design.

4.4.2 Radiometry in Optical Systems

These ideas can be applied to real optical systems. Now, however, for purposes of simplification, the cosine projections can be dropped. (This is certainly a good approximation for the source, since the distances are so great and the angles so small. It is also good for the detector, which is mounted normal to the optical axis.) There are two cases of interest to be considered. One is for the so-called point source; the other is for an extended source. The *point source* is a source that is not resolved. It has an area smaller than the entrance window. It does have a finite area, or it would not radiate, but no details of the area are available to the optical system. An extended source, on the other hand, is a source that is always larger than the image of the field stop or detector on it. These are both shown in Fig. 4-5.

The fundamental theorem of transfer can be applied to the point source. (This is the small source shown in Fig. 4-5.) The power on the entrance pupil is the source radiance times the source area times the aperture area divided by the square of the range.

$$\Phi = \tau_a L \frac{A_s A_o}{R^2} = \tau_a L \Omega_{so} A_o = \tau_a L A_s \Omega_{os} , \qquad (4.9)$$

where the subscript s stands for *source*, o stands for *optics*, d stands for *detector*, and τ_a is the atmospheric transmission. In the case of a solid angle, a subscript *so* means the *source as seen from the optics*. The power can also be written as the radiance times the source area times the solid angle the aperture subtends at the source. The only way the

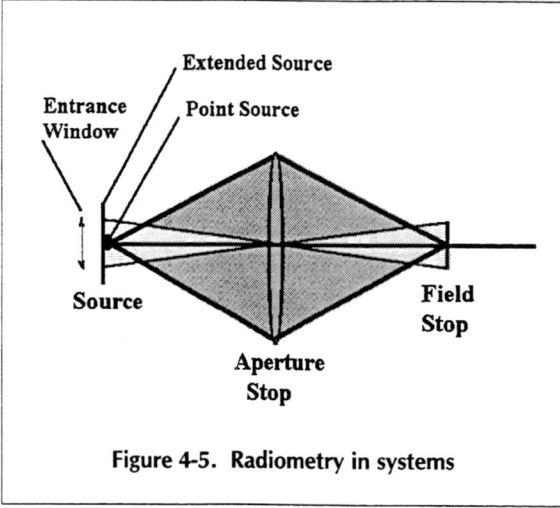

Figure 4-5. Radiometry in systems

power on the aperture can be increased for a fixed source is by a decrease in range or by an increase in aperture area. Both increase the solid angle of collection. The source radiance times its area is the intensity, and this is usually the way such a source is described. The power on the detector is the same as that on the aperture, but the flux densities are different by the ratio of their respective areas. This ratio is sometimes called the optical gain.

$$E_d = \tau_a \tau_o \frac{A_s A_o}{A_d R^2} = \tau_a \frac{\Phi A_0}{A_0 A_d} = \tau_a E_0 G_o \ . \tag{4.10}$$

Extended sources are different. The effective source area is the image of the detector on the larger, extended source. Thus there is a similar-triangle relationship, and the power on the detector can be written in terms of the source variables and in terms of the sensor variables. There are two different kinds of solid angles involved, both of which can be used for the radiometric calculations, but it is useful to think of them as resolution angles and radiometric angles. The resolution angle in this case is the angle formed by the detector at the aperture, and (equivalently) the image of the detector formed at the object distance. These alternate angles are shown in Fig. 4-6. The top of the figure shows resolution half angles; the bottom shows speed half angles.

The radiometric angles are the angle the aperture subtends at the source and at the detector, Ω_{os} and Ω_{od}. The speed of the optical system is written in terms of the radiometric solid angle on the image side. This solid angle can be written either as $\pi \sin^2 \theta$, as we saw above, or as $\pi/(4F^2 + 1)$, where F is the F number. For a well-corrected system the solid angle is $\pi/4F^2$. The speed is also given in terms of the numerical aperture NA.

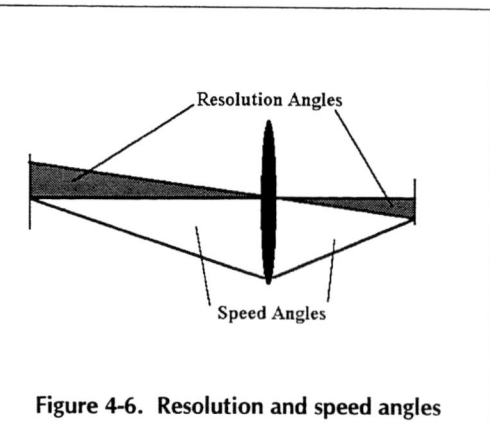

Figure 4-6. Resolution and speed angles

$$P = \tau_o LA_d \Omega_{od} = \tau_o LA_d \frac{\pi}{4F^2} = \tau_o LA_d \pi NA^2 . \tag{4.11}$$

An alternative is in terms of the area of the aperture and the resolution angle.

$$P = \tau_o LA_o \frac{A_d}{f^2} = \tau_o LA_o \alpha^2 , \tag{4.12}$$

where α is the angular subtense of one side of the (square) detector (and α^2 is A_d/f^2).

4.4.3 Reflection, Transmission, Absorption, and Emission

Emissivity and absorptivity are defined as the ratios of the flux absorbed or emitted to that of a perfect radiator or a perfect absorber; they are emission and absorption efficiencies. Reflection and transmission are the ratios of radiation that is reflected or transmitted from a sample to the radiation that is incident. Emission and absorption are spectrally varying quantities and can therefore be described as spectral, total, or band limited. That is, they can be specified as a set of values at a set of wavelengths, as an average over the entire spectral range from zero to infinity, or over a finite, specific spectral interval. The spectral value is

$$\epsilon(\lambda) = \frac{R(\lambda)}{R^{BB}(\lambda)} , \tag{4.13}$$

where $R(\lambda)$ is some radiometric function—either radiance or radiant exitance (emittance)—and the superscript *BB* represents a blackbody. In truth, the only value that can be measured is the weighted average over a finite band. This can be represented as follows:

$$\epsilon_{h,1\Delta\lambda} = \overline{\epsilon_h} = \frac{M_{\Delta\lambda}}{M_{\Delta\lambda}^{BB}} = \frac{\int_{\Delta\lambda} \epsilon(\lambda) M^{BB}(\lambda,T) d\lambda}{\int_{\Delta\lambda} M^{BB}(\lambda,T) d\lambda} . \tag{4.14}$$

If the spectral band $\Delta\lambda$ is from zero to infinity, then this is called a total emissivity. The total, hemispherical emissivity is defined as

$$\epsilon_{t,h} = \frac{M}{M^{BB}} . \tag{4.15}$$

The total, directional emissivity is defined as

$$\epsilon(\theta,\phi) = \frac{L(\theta,\phi)}{L^{BB}} .$$

(4.16)

Other definitions for the other spectral definitions (spectral and weighted average) are obvious. Thus there are six different types of emissivity and six types of absorptivity—spectral, average, and total, combined with directional and hemispherical.

Reflection and transmission are functions of both the input and the output geometries; they are bidirectional quantities in general, depending upon both input and output directions. They differ only in geometry—whether the beam goes through the sample or is reflected back from it. Specular reflectivity and transmissivity are dimensionless in that they are ratios of flux or flux densities in the (specular) beam.

Transmittance and reflectance, however, can be directional, for a specific direction, or hemispherical, the average value over the entire overlying hemisphere. The definitions of both hemispherical reflectivity and transmissivity are

$$\tau_{dh} = \frac{M}{E(\theta,\phi)} \quad \rho_{dh} = \frac{M}{E(\theta,\phi)} .$$

(4.17)

The geometry dictates the difference. In addition to specular quantities for transmission and reflection, hemispherical quantities may also be specified, but now there exist two directions, the input and the output, and such double adjectives as directional-hemispherical, hemispherical-hemispherical, and hemispherical-directional all arise. The more fundamental and useful quantity for samples that have some degree of scatter is the so-called bidirectional reflectance (transmittance) distribution function (BRDF or BTDF). This is the ratio of output radiance to input incidence (irradiance).

$$\rho_b = \frac{L^{ref}(\theta',\phi')}{E^{inc}(\theta^i,\phi^i)} .$$

(4.18)

It is dimensionless, but has the units of reciprocal steradians. It has a maximum theoretical value of infinity, maximum practical value of about 10^4 sr^{-1}, and has a value of the hemispherical reflectivity divided by π for an isotropic sample. Its main utility is that for a given geometry, in which the incidence (irradiance) on the sample may be calculated, the output radiance is then available to be inserted in the fundamental equation of transfer to get the next transfer.

Kirchhoff's law relates emissivity to absorptivity. For a body in equilibrium, and under the same conditions of temperature, spectrum, and the like, the absorptivity of a sample is equal to its emissivity—totally, spectrally, for each component of polarization and in each direction. It is not true, however, that for an opaque sample the emissivity is one minus the reflectivity. One counterargument is simply that the geometries are different. It can be shown, however, that for the opaque sample the hemispherical emissivity is the ones complement of the hemispherical reflectivity. This is also true for specular reflectivity of specular samples.

4.4.4 Blackbody (Perfect) Radiators

The blackbody radiator is called that because a perfect absorber is a perfect emitter. Thus, blackbodies may be thought of as bodies that have 100% emissivity. They radiate all the flux density at every wavelength that their equilibrium temperature allows. In 1900 Planck first established the radiation laws for these bodies, and at the same time introduced quantum mechanics to the world. The radiant exitance (emittance) in $Wcm^{-2}\mu m^{-1}$ emitted by a blackbody may be expressed as

$$M_\lambda = c_1\lambda^{-5}(e^x-1)^{-1} , \tag{4.19}$$

where c_1 is the first radiation constant, given by $2\pi c^2 h$, and x is the dimensionless frequency given by $x = hc/\lambda kT$, where c is the velocity of light in vacuum, h is Planck's constant, k is Boltzmann's constant, and T is the temperature in absolute units, kelvins preferred. Figure 4-7 shows blackbody curves in terms of the radiant exitance (emittance) as a function of wavelength for a number of different temperatures, each one of higher temperature completely above the others. The wavelength of the peaks can be found from the distribution law

$$\lambda_{max}T = 2898\,[\mu mK] . \tag{4.20}$$

The peak wavelength in micrometers is approximately 3000 divided by the temperature in kelvins. The same curve can be integrated to obtain the expression for total radiation in all wavelengths:

$$M = \sigma T^4 , \tag{4.21}$$

where σ is the Stefan Boltzmann constant (5.67×10^{-8} $Wm^{-2}K^{-4}$). The expression can also be expressed in terms of other spectral variables, notably wave number and dimensionless frequency:

$$M_\nu = 2\pi hc^2\sigma^4(e^x-1)^{-1} . \tag{4.22}$$

$$M_x - 2\pi c^2 h x^3(e^x-1)^{-1} . \tag{4.23}$$

The curve of radiant exitance (emittance) as a function of wave number is shown as Fig. 4-7. Each of these exitances (emittances) may also be expressed as a photon rate by dividing by the energy of a photon hc/λ. The peaks and distributions of all of these are different, but the total power density is the same. The total photon density is not the same as the total power density.

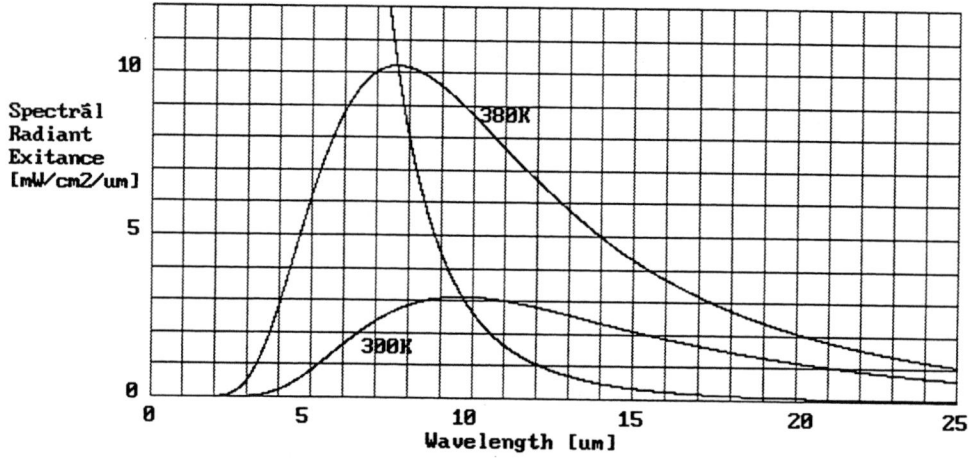

Figure 4-7. Spectral radiant exitance vs wavelength

None of them integrate analytically. Programs are provided in Appendix B for generating these figures and for making calculations of the integrals over any desired spectral band. Programs for the logarithmic versions are also shown. There are also normalized so-called universal curves. These are essentially of two types, distribution and cumulative functions. The first are plots with the dimensionless frequency $x = c_2/\lambda T$ as the abscissa and the exitance (emittance) normalized to the maximum as the ordinate. The cumulative versions are plots of the integral from 0 to x as a function of x, normalized to the appropriate Stefan Boltzmann expression for the total radiation.

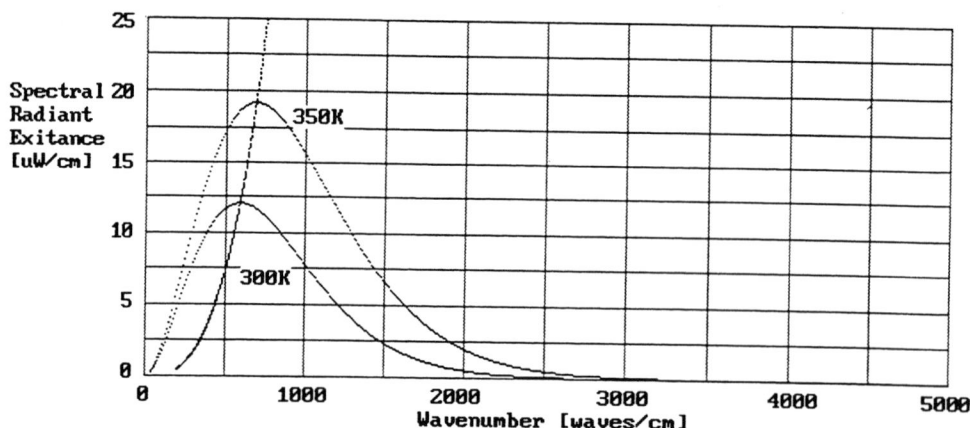

Figure 4-8. Spectral radiant exitance vs wave number

5 DETECTOR PARAMETERS

The detector is often called the heart of the infrared system. This is because the system itself is so dependent on the performance of the detector. In this chapter the ways in which detectors are described, and the different types and some properties of arrays, are discussed. Details of performance are left for other texts. See Appendix A.

5.1 Detector Descriptions

The property of infrared detectors as transducers is usually described in terms of their responsivity. This may be given in volts per watt, amps per watt, electrons per watt second, or some equivalent terms. The ratio is always that of an output voltage or equivalent to an input radiant power or equivalent. These are not square-law detectors; they have a linear relationship between the radiant power in the input signal and the voltage or current in an output electrical signal. For illustrative purposes and convenience in exposition, I will assume the input is power and the output is a voltage.

$$V_s = \int \Re(\lambda)\Phi(\lambda)\,\mathrm{d}\lambda \ . \tag{5.1}$$

This responsivity $\Re(...)$ is a function of many things; the principal input parameters that cause it to change are spectral distribution, temporal distribution, spatial distribution, and to a lesser extent, input signal level. They are linear, shift-invariant systems to only a small degree. The operating parameter of most importance is the bias level.

The noise of a detector is also dependent on very many things. The noises may be Johnson, shot, generation-recombination, photon, temperature, or excess, and the preamplifier noise has an important role to play as well. No matter—the total noise is the root mean square sum of each of the individual noise components. The total noise is obtained by integrating the noise spectrum $V_n(f)$ over the frequencies f where there is sensitivity. However, it is done on a quadrature basis.

$$V_n^2 = \sum_i \int V_{ni}^2(f)\,\mathrm{d}f \ , \tag{5.2}$$

where the summation is over the individual noises, described below, and the integration is over the frequency range of sensitivity of the detector. The output signal-to-noise ratio may therefore be written as the input power times the responsivity divided by the noise.

$$\mathrm{SNR} = \frac{V_s}{V_n} = \frac{\int\!\int \Re(\lambda,f)\Phi(\lambda,f)\mathrm{d}\lambda\,\mathrm{d}f}{\left(\sum \int V_n^2 \mathrm{d}f\right)^{\frac{1}{2}}} \ . \tag{5.3}$$

This signal-to- noise ratio (SNR) is a function not only of the type of noise and inherent properties of the detector, but it is also a function of the input power, the detector size and the bandwidth of the noise spectrum. In comparing different detectors, it is useful to normalize some of these properties. One way to do this is to define the Noise Equivalent Power or NEP of a detector. It is the input signal power that gives a signal-to-noise ratio of one. It is, said a different way, the signal that is just equal to the noise. This is still a function of the detector size and the noise bandwidth.

$$\text{NEP} = \frac{V_n}{\Re(\lambda, f)} \, .$$

(5.4)

It is also unfortunate that the NEP is better when it is lower, and this is un-American. Thus, detectivity was defined as the reciprocal of the NEP.

$$D = \text{Detectivity} = \frac{1}{\text{NEP}} = \frac{\text{SNR}}{\Phi} \, .$$

(5.5)

The specific detectivity was also defined as the detectivity for a 1-cm^2 detector using a 1-Hz noise bandwidth. Detectors are almost never used this way, but they can be compared on this basis. The specific detectivity is

$$D^*(\lambda, f, \Delta f) = \frac{\sqrt{A_d B}}{\Phi} \text{SNR} \, ,$$

(5.6)

where the arguments of D* are the wavelength, the modulation frequency, and the bandwidth. The last-named must be one by definition. The modulation frequency helps to indicate whether it is in the region of $1/f$ noise or not.

It is useful to have the expression for an idealized detector, one that is limited only by the noise in the incoming photon stream. It is the so-called background-limited infrared photodetector, or BLIP detector. You can't do better, and it is often useful to find out how well you can do. The BLIP D* is given by

$$D^*_{\text{BLIP}} = \frac{\lambda_m}{hc} \sqrt{\frac{\eta}{gE_q}} \, ,$$

(5.7)

where D*$_{\text{BLIP}}$ means the value at the maximum, λ_m is the wavelength of the maximum, h and c are the Planck constant and speed of light, η is the quantum efficiency (the number of electrons generated for each incident photon), g is 2 for detectors without recombination and 4 for those with it, and E_q is the photon flux density on the detector.

5.2 Detector Noises

Different types of noises can adversely affect the performance of detectors. These include photon noise and temperature noise as fundamental noise limits, and Johnson, shot, generation-recombination, and excess as internal noises.

Photon noise can be considered as shot noise in the photon stream. For all practical purposes it follows Poisson statistics, and therefore the mean square fluctuation from the mean rate of the photon stream is equal to the mean rate.

Temperature noise is the fluctuation in the temperature of a thermal detector due to the fluctuations in the emitted and absorbed powers (due to the fluctuations in photon rates), and the D* is given by

$$D^* = \frac{\epsilon}{\sqrt{8\sigma k (T_s^5 + T_d^5)}} \, , \tag{5.8}$$

where σ is the Stefan-Boltzmann constant, k is the Boltzmann constant, ϵ is the emissivity of the detector, and the two temperatures are those of the detector and the scene.

Johnson noise, alternately called thermal noise, and not to be confused with temperature noise described above, is due to the fluctuations in the flow of carriers in a resistance. Its mean square current is given by

$$i_{ms} = \frac{4kT}{R}B \, , \tag{5.9}$$

where k is the Boltzmann constant, T is the detector temperature, R is the detector resistance, and B is the bandwidth.

Shot (Schottky) noise is a result of the fluctuation in the rate at which carriers surmount a potential barrier. Its mean square current is given by

$$i_{ms} = 2qIB \, , \tag{5.10}$$

where q is the charge on the electron, I is the average (or dc) current, and B is the bandwidth.

Generation-recombination noise is due to the fluctuations in the rate at which carriers are generated and recombine. It is given by

$$i_{ms} = \frac{I4\tau}{\overline{N}(1 + \omega^2\tau^2)}B \, , \tag{5.11}$$

where τ is the recombination time and $\overline{N}$ is average photon number.

Excess noise, also called $1/f$ noise, modulation noise, and current noise, is not well understood, but does depend on the quality of contacts and surface passivation. It is given by

$$i_{ms} = \frac{KI^\alpha}{f^\beta}B \, , \tag{5.12}$$

where K is a proportionality constant dependent upon the detector, α is a constant close to 2, and β is a constant approximately equal to 1. That really is empirical!

These noises are shown asymptotically and in an idealized form in the accompanying log-log plot, Fig. 5-1. There is no significance in the magnitude of any of them, for they change positions with different detectors and different flux densities.

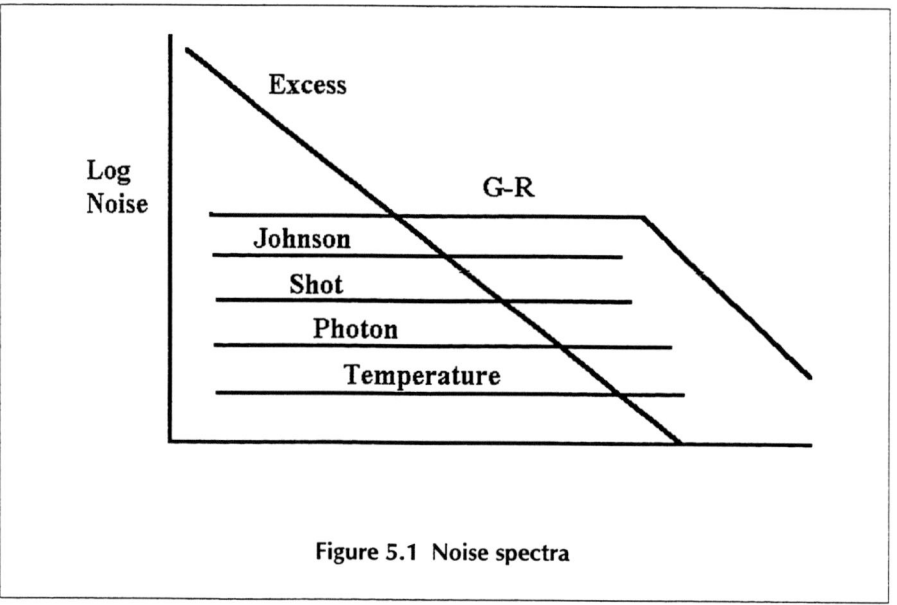

Figure 5.1 Noise spectra

5.3 Detector Types

Detectors can be categorized in two different ways: by the mechanisms of their transduction and by their areal characteristics. The first of these divides detectors into thermal and quantum detectors. Thermal detectors respond to their own temperature or temperature change. Thus, an input optical beam power is absorbed; the heat generated over a short period of time gives rise to an increase in temperature of the detector material, and some physical property of the detector that changes with this temperature change is monitored. If the effect is the change in resistance with temperature, the detector is called a bolometer. There are several kinds: metal, superconducting, and semiconducting. If the change is the contact difference in potential, the detector is called a thermocouple or thermopile. If it is the internal electric (polarization) field, the detector is called pyroelectric. Less popular are gas-expansion or pneumatic detectors, and linear expansion versions.

If the detector responds directly to the incidence of photons—for example, if there is one electron excited to the conduction band for every two photons (on average)—the detector is called a photodetector or photon detector. If the resistance (or conductance) of the detector is monitored, it is a photoconductor. If the dipole voltage is monitored, it is called a photovoltaic detector if no bias is applied and a photodiode if bias is applied. Other effects, using magnetic fields, the photoelectromagnetic effect, and shifts of the conduction band (Dember effect), are sometimes used. The details of the operation of detectors are varied and fascinating.

An ideal thermal detector has a responsivity that does not vary with wavelength. Most thermal detectors approximate this fairly well. An ideal photon detector responds equally to all photons in its spectral band. On an energy basis, however, the response of a photon detector increases linearly with wavelength until its cutoff wavelength. The response is equal for all wavelengths in terms of electrons per photon, but longer wavelength photons have less energy, so the response in energetic terms is higher for the longer-wavelength photons.

Thermal detectors have a specific detectivity of about 10^8 to 10^9 cm Hz$^{1/2}$W^{-1} and time constants of about 1 msec. Photon detectors are usually photon-noise limited at values up to about 10^{13} with time constants less than 1 μs. Different detector types are listed in Table 5-1.

Table 5-1. Types of Detectors

Photon
Photodiode
Photoconductive
Photoelectromagnetic
Thermal
Bolometers
Metal
Superconducting
Semiconducting
Thermocouples
Thermopiles
Pyroelectrics
Expansion
Golay

5.4 Detector Arrays

Early infrared imaging devices used a single detector and scanned it over the field of view, registering the signal on a point-by-point basis. As a result of progress in the semiconductor industry, and in particular with charge-coupled devices, there are now two-dimensional detector arrays that can accomplish this function without scanning. Although not all imaging functions can now be accomplished by such staring arrays, they have become extremely important in infrared imaging. Accordingly, some properties are reviewed here.

There are two types of arrays—hybrid and monolithic. In the former, the detection process is carried out in the chosen material; the electronic charges are then transferred to a shift-register system, usually silicon, by way of indium interconnections in the forms of little bumps of indium. In a monolithic array, currently limited to doped-silicon detectors, both detection and charge manipulation are done in the one material.

Table 5-2 summarizes the different types of arrays available in the different materials. Since development is intense as well as rapid in this field, the numbers can only be considered a guide. Some manufacturers have probably already exceeded some of these values.

It is shown later that some of the important properties of detector arrays are the number of elements, element areas, quantum efficiencies, well size, and uniformity. In some applications the speed of response and clock times need to be evaluated as well.

Indium antimonide is important for the MWIR and is a relatively mature technology. Mercury cadmium telluride is applicable to both the MWIR and LWIR, but it is more difficult to manufacture for the LWIR than the MWIR.

Doped silicon is a good LWIR detector, but it must be cooled to rather low temperatures, on the order of 20K.

Table 5-2. Array Properties

Material	Spectrum, μm	Count	QE	Size	Type
Doped Si	3–5 & 8–24	512×512	0.2	10^6	Mono
PtSi	3–5	1024×1024	0.01	10^5	Hybrid
HgCdTe	3–5 & 8–12	256×256 1028×4	0.8	10^6	Hybrid
InSb	3–5	256×256	0.8	10^6	Hybrid
PbSe	3–5	128×128	0.8	?	Hybrid
Thermal	3–5 & 8–12	250×350			Hybrid
Si	0.3–1.1	2000×2000	0.9	10^6	Mono

Platinum silicide is very uniform and therefore attractive for many applications, even though it has a very low quantum efficiency. Apparently because of its uniformity and material parameters, it has the largest format, i.e., number of pixels in the array. The bolometric and pyroelectric arrays need not be cooled and are attractive for those applications where cooling is anathema but the ultimate in sensitivity is not necessary. They are made by etching holes in a silicon substrate and depositing the detector material. The elements could be considered "Bridges on the River Silicon." Two types of bolometers are available: one is a pyroelectric (or ferroelectric) array; the other is a thermistor bolometer. They are neither monolithic nor do they have bump bonds. They have detectivity values of about 10^9, but that is sufficient to make systems with detectable temperature differences of about 0.1K. (The cooled detectors have detectivities in the terrestrial environment of about 10^{11}.)

6 THE INFRARED SYSTEM

The design and analysis of infrared systems requires a consideration of the targets, the background, and the intervening medium, as well as all the components of the equipment itself, the optics, the detectors, the electronics, and the display or processing.

This chapter, a rather mathematical one, derives some simple expressions for the SNR in terms of the target, background, transmission, and components. Although the results have certain approximations and idealizations, they are a useful starting point in the iterative design process. The steps in the derivation are just as important. The observant reader can then adapt the processes to his own application.

6.1 Description of the Infrared System

The infrared system consists of the target, the background, the properties of the intervening medium, the optical system, the detector (transducer), the electronics, and the display. These are all shown in Fig.6-1. Every infrared system has all of these, although some may be much more elaborate than others. It is tempting to think of the system as just the equipment and not the source, background, and medium. Yielding to this temptation is a sin that will be redeemed in failure. All these components, substances, and physical entities must be considered. The target is what you want to see; the background is what you don't. One man's target is another man's background. Consider the case of people walking in front of a cornfield. If you are designing a remote sensing system, the cornfield is the target and the people are background, even though they are in the foreground. If you are doing night driving, the people are the target(s) and the cornfield is the background. Note that sometimes targets are to be hit and sometimes they are to be avoided. The atmosphere may reduce the level from both the target and background by transmission losses that arise from both absorption and scattering. The atmosphere may increase the nonsignal level on the sensor by scattering light from out

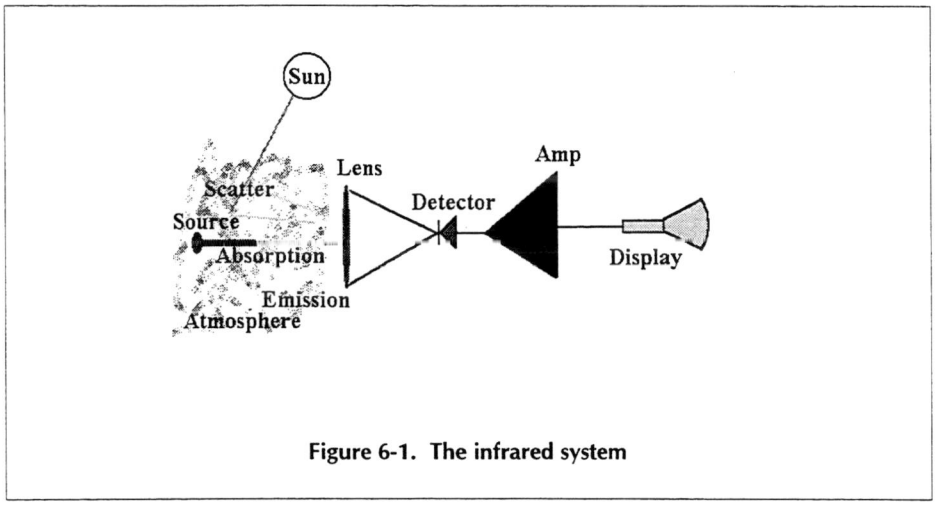

Figure 6-1. The infrared system

of the field of view into it. The atmosphere may increase the level of nonsignal radiation by its own emission. All the components of radiation that are not attributed to the target are lumped into the term "background." So some background can come from the foreground!

These sources of radiation and attenuation will be covered in subsequent chapters; the mathematical representation of signal-to-noise ratio is considered in this chapter.

6.2 Sensitivity Equations

There is one very basic signal-to-noise equation. It uses the signal, calculated as the responsivity $\Re$ times the power on the detector P_d, for the numerator. The product of the spectral distribution of these quantities is integrated over the optical spectrum and the electrical one. The noise is the square root of the integral of the square of the noise spectrum.

$$\text{SNR} = \frac{\int\int \Re(\lambda, f)\Phi_{\lambda, d}(\lambda, f)\mathrm{d}\lambda \, \mathrm{d}f}{\sqrt{\int V_n^2(f)\mathrm{d}f}} \, . \tag{6.1}$$

In extreme cases, it may be necessary to integrate the power over the angles of acceptance and over the area of the detector using small enough solid angles and small enough detector areas that the power is constant over them. Every detector has a response that is a function of its area (its sensitivity contour) and a function of incidence angle. Some are sufficiently uniform that this need not be considered. Equation (6.1) is fundamental and may be necessary in the most detailed of calculations. But there are other forms that employ certain simplifying assumptions that are practical and useful, and we shall cover some of them in this section. However, if there is any doubt, and the consequences are severe, calculate from this fundamental relationship.

If the detector can be described accurately by its specific detectivity, then for a given wavelength

$$D^*(\lambda, f, B) = \frac{\sqrt{A_d B}}{\Phi_{\lambda, d}(\lambda)} \, \text{SNR} \, , \tag{6.2}$$

and over some spectral band, the SNR is given by

$$\text{SNR} = \frac{D^*\Phi_d}{\sqrt{A_d B}} = \frac{1}{\sqrt{A_d B}}\int D^*(\lambda)\Phi_{\lambda, d}(\lambda)\mathrm{d}\lambda \, . \tag{6.3}$$

The expressions for SNR for the case of point sources and extended sources, using the simplifying assumption that the detectors may be described adequately by the specific detectivity, may now be written.

6.2.1 Point Sources

The sensitivity, in this case the SNR, can be calculated using the definition of specific detectivity and the fundamental equation of transfer for a point source. After manipulating the nonspectral variables, we can obtain an appropriate equation for the SNR from a point source.

$$\text{SNR} = \frac{A_s A_o}{R^2 \sqrt{A_d B}} \int \tau_a \tau_o L D^* d\lambda , \qquad (6.4)$$

where the arguments of the variables inside the integral have been suppressed, the integration is assumed to be over the spectral region of interest, and the subscripts on the spectral distributions have also been suppressed. These simplifications will be continued unless there is reason to show them explicitly. The real radiance may be rewritten as the emissivity times the blackbody radiance:

$$\text{SNR} = \frac{A_s A_o}{R^2 \sqrt{A_d B}} \int \tau_a \tau_o \epsilon L^{BB} D^* d\lambda , \qquad (6.5)$$

$$\text{SNR} = \frac{\bar{\epsilon} A_s A_o}{R^2 \sqrt{A_d B}} \int \tau_a \tau_o L^{BB} D^* d\lambda . \qquad (6.6)$$

The emissivity was pulled out of the integral by writing it as the weighted average emissivity:

$$\bar{\epsilon} = \frac{\int \epsilon \tau_a \tau_o L^{BB} D^* d\lambda}{\int \tau_a \tau_o L^{BB} D^* d\lambda} . \qquad (6.7)$$

In these equations, A_s = area of the source, A_o = area of the optical aperture, A_d = area of the detector, R = range, D^* = specific detectivity, B = effective noise bandwidth, τ_a = atmospheric transmission, τ_o = optical transmission, L = radiance, L^{BB} is blackbody radiance, and ϵ is directional emissivity. The effective noise bandwidth is described in Sec. 9.5. For now, assume that it is bandwidth that extends from dc to the highest required frequency. In this form, the average emissivity and the source area are indicated. The solid angle of the optics subtended at the source is next, and it involves the inverse square of the range and the optical area. The detector area and bandwidth are in the denominator. Inside the integral are the two transmission factors. Atmospheric transmission is a function of the range and other geometries, as we shall see. We have reduced the source problem to that of a blackbody by using the weighted, average emissivity. Finally there is the D^*. This form of the integral will be seen again.

Equation (6.6) is a variety of the equation that is usually used in the formation of ballistic missile detection problems, infrared search and track problems, early warning and tracking systems, but it is not limited to those applications.

Inspection of the equation gives information on how to create a design. Starting from the first terms on the right side of Eq. (6.6), it is clear that the emissivity-area product ϵA_d should be as large as possible and the range as short as possible. These are usually not design variables, but standoff requirements and properties of the target. Typical values of the emissivity-area product for a reentry vehicle range from 0.1 to 1 m^2. The designer *can* make the detector as small as possible, at least as small as technology dictates. This is usually on the order of 50 to 25 μm. Sometimes this is limited by some of the optical characteristics of the systems. For example, it better not be smaller than the image blur of a point source. The designer *can also* make the optics as large as possible, but 1 m is usually the maximum that can be put in a vehicle. The bandwidth will be determined by the frame rate required for repeated looks, the total coverage, and the number of detectors used in the system. The atmospheric transmission can be maximized by keeping the entire path above the sensible atmosphere. That makes it unity. If this is not permitted by the geometry, that is, the altitude of the sensor package, the altitude of the target, and the required range, then all that can be done is to calculate the transmission by the methods discussed in the next chapter. The radiance of the target is calculated straightforwardly based on its temperature which ranges from 200 K to 300 K. So it's all there. Just do it!

6.2.2 Point Sources with Photon-Limited Detectors

The following equation is the same as that derived earlier; it is just repeated here for convenience.

$$\text{SNR} = \frac{\overline{\epsilon} A_s A_o}{R^2 \sqrt{A_d B}} \int \tau_a \tau_o L^{BB} D^* d\lambda \ . \tag{6.8}$$

Now if we pull out the transmission terms by defining effective transmittances as indicated, then we can get to the heart of the spectrum. Note, we have already defined an effective emissivity in Eq. (6.7). Now, the effective atmospheric transmission is also a weighted average, but it does not have the emissivity as part of the weighting function.

$$\text{SNR} = \frac{A_s A_o}{R^2 \sqrt{A_d B}} \overline{\epsilon} \, \overline{\tau}_a \overline{\tau}_o \int L^{BB} D^* d\lambda .$$ (6.9)

Then, the effective optical transmittance is a weighted average with neither the emissivity nor the atmospheric transmission as part of the weight.

$$\overline{\tau}_a = \frac{\int \tau_a \tau_o L^{BB} D^* d\lambda}{\int \tau_o L^{BB} D^* d\lambda} \qquad \overline{\tau}_o = \frac{\int \tau_o L^{BB} D^* d\lambda}{\int L^{BB} D^* d\lambda} .$$ (6.10)

We could have done it the other way around, but this seems to make sense. They are arbitrary, but consistent and correct definitions. In many cases these factors will be flat over the spectrum of interest. The emphasis here is with the rest of the integral, the integral of the product of the specific detectivity and the radiance. If the detector is an idealized photon detector, one with a quantum efficiency that is independent of wavelength, then the following linear relationship is valid.

$$D^*(\lambda) = \frac{\lambda}{\lambda_{max}} D^*_{max} .$$ (6.11)

A little substitution shows that the integral can be rewritten with the photon radiance. The photon radiance (photance) is related to the energetic radiance (radiance) by the energy of a photon at a given wavelength

$$L^{BB} = \frac{hc}{\lambda} L_q^{BB}$$ (6.12)

and

$$\int D^* L^{BB} d\lambda = \int (D^*_m/\lambda_m) \lambda L^{BB} d\lambda$$ (6.13)

$$= (D^*_m/\lambda_m) \int \lambda L^{BB} d\lambda = D^*_m \frac{hc}{\lambda_m} \int L_q^{BB} d\lambda .$$ (6.14)

Further, if the detector is photon limited, (the noise is generated predominately, if not entirely by the fluctuation in the arrival rate of the photon flux at the detector), then the

D^* at the maximum is given by the following equation. This may also be considered the result of the shot noise in the photon stream.

$$D_m^* = D_{\text{BLIP}}^* = \frac{\lambda_m}{hc}\sqrt{\frac{\eta}{gE_q}} \tag{6.15}$$

and the final equation for SNR devolves to the effective emissivity, the effective transmittances, the quantum efficiency, a geometric term for collecting signal flux, and the noise bandwidth.

$$\text{SNR} = \overline{\varepsilon\tau_a\tau_o}\eta^{1/2}\,\frac{A_S A_O}{R_2\sqrt{A_d B}}\,\frac{\int L_q^{BB}d\lambda}{\sqrt{\int E_q\,d\lambda}} \tag{6.16}$$

The range is also in there, and a big factor it usually turns out to be. The final term is the ratio of the blackbody radiance in the band from the target to irradiance on the detector from all sources.

The range and emissivity-area product are there as in Eq. (6.9). One should still maximize the optics size and minimize the detector size. In addition, however, one should maximize the quantum efficiency and minimize the background flux E_q that gets to the detector. Since $E = L\,A_o/f^2$, the area of the optics is less important and the focal length should be made large—to reduce the extended background flux.

6.2.3 Extended Source Temperature Sensing

Usually, extended source detections and imaging have to do with temperature sensing, or at least the difference in temperature between one part of the scene and another. The equation derived earlier for the extended source case was

$$\text{SNR} = \frac{A_o A_d}{f^2\sqrt{A_d B}}\int \tau_a\tau_o LD^*d\lambda \ . \tag{6.17}$$

The transmissivity and emissivity can be pulled out of the integral as before:

$$\text{SNR} = \frac{A_o A_d\overline{\tau_a}\,\overline{\tau_o}\,\overline{\epsilon}}{f^2\sqrt{A_d B}}\int LD^*d\lambda \ . \tag{6.18}$$

A little geometry and some algebra can be used to get:

$$\text{SNR} = \frac{\pi D^2}{4f} \frac{\sqrt{A_d}}{f} \frac{\overline{\tau_a \tau_o \epsilon}}{\sqrt{B}} \int L^{BB} D^* d\lambda \ , \tag{6.19}$$

$$\text{SNR} = \frac{\pi}{4} \frac{D\alpha}{F} \frac{\overline{\tau_a \tau_o \epsilon}}{\sqrt{B}} \int L^{BB} D^* d\lambda \ . \tag{6.20}$$

This is the signal-to-noise ratio in terms of the specific detectivity; it can be adjusted for a photon-noise-limited detector. The expression for that D^* is

$$D_{\text{BLIP}}^* = \frac{\lambda_m}{hc} \sqrt{\frac{\eta}{g E_q}} \ . \tag{6.21}$$

where E_q is the total photon incidance on the detector. So now the equation can be written

$$\text{SNR} = \frac{\pi}{4} \frac{D\alpha}{F} \frac{\overline{\tau_a \tau_o \overline{\epsilon}}}{\sqrt{B}} \lambda_m \sqrt{\frac{\eta}{g}} \frac{\int L_q^{BB} d\lambda}{\sqrt{\int E_q(\lambda) d\lambda}} \ . \tag{6.22}$$

where the integral of E_q is written explicitly for emphasis. Now, because the incidance is related to the radiance by

$$E = L\Omega = L\frac{\pi}{4F^2} \ , \tag{6.23}$$

then

$$\text{SNR} = \frac{\sqrt{\pi}}{2} D\alpha \frac{\overline{\tau_a \tau_o \overline{\epsilon}}}{\sqrt{B}} \sqrt{\frac{\eta}{g}} \frac{\int L_q^{BB} d\lambda}{\sqrt{\int L_q(\lambda) d\lambda}} \ . \tag{6.24}$$

The one additional step toward idealization is to assume that the optical system is diffraction limited. In this case, based on placing the central diffraction disk on the detector, one has

$$D\alpha = 2.44\lambda \ . \tag{6.25}$$

Therefore

$$\text{SNR} = 0.84\frac{\sqrt{\pi}}{2}\bar{\lambda}\frac{\bar{\tau}_a\bar{\tau}_o\bar{\epsilon}}{\sqrt{B}}\sqrt{\frac{\eta}{g}}\frac{\int L_q^{BB}d\lambda}{\sqrt{\int L_q(\lambda)d\lambda}} \ . \tag{6.26}$$

The 0.84 factor arises from the fact that 84% of the total diffraction pattern is in the central lobe of the diffraction pattern. Evaluation of the numbers yields

$$\text{SNR} = \frac{0.74\bar{\lambda}\bar{\tau}_a\bar{\tau}_o\bar{\epsilon}}{\sqrt{B}}\sqrt{\frac{\eta}{g}}\frac{\int L_q^{BB}d\lambda}{\sqrt{\int L_q(\lambda)d\lambda}} \ . \tag{6.27}$$

This interesting form shows that the SNR is equal to several efficiency factors times the ratio of the photance of the signal to the square root of the photance of the total radiation on the detector, the square root of the bandwidth, and directly to the average wavelength. The final, idealized SNR equation is obtained when all the efficiency factors are assumed to be unity. Then

$$\text{SNR} = \frac{0.74\bar{\lambda}}{\sqrt{B}}\left[\frac{\int L_q^{BB}d\lambda}{\sqrt{\int L_q(\lambda)\,d\lambda}}\right] \ . \tag{6.28}$$

In this form it is clear that the SNR is a function of the input signal-to-noise ratio of the radiation field, the term in the square brackets, the information bandwidth, and the wavelength that determines the diffraction limit.

Now it makes sense to derive the expression for the minimum temperature difference that can be found. This, of course, is the NETD. The NETD is the temperature difference that gives a signal that is just equal to the noise. It is also the ratio of the temperature difference to the change in the SNR. Thus, its reciprocal is the derivative of the SNR with respect to the temperature. Using this, one has

$$\frac{1}{\text{NETD}} = \frac{\partial \text{SNR}}{\partial T} = \frac{0.74\bar{\lambda}\bar{\tau}_a\bar{\tau}_o\bar{\epsilon}}{\sqrt{B}}\sqrt{\frac{\eta}{g}}\frac{\int \frac{\partial L_q^{BB}}{\partial T}d\lambda}{\sqrt{\int L_q(\lambda)\,d\lambda}} \ , \tag{6.29}$$

an idealized equation with many efficiency factors included in it. It represents the NETD for a photon-limited, diffraction-limited, circularly symmetric system that uses a detector that can be described by the specific detectivity figure of merit. The completely idealized form, in which all the efficiency factors are one, is

$$\frac{1}{\text{NETD}} = \frac{\partial \text{SNR}}{\partial T} = \frac{0.74\overline{\lambda} \int \dfrac{\partial L_q^{BB}}{\partial T} \, d\lambda}{\sqrt{B} \, \sqrt{\int L_q(\lambda) \, d\lambda}} \, . \tag{6.30}$$

6.2.4 Charge-Collection Devices

The modern-day era of charge-collection types of detectors, whether CCD structures or special detectors with capacitive storage, has ushered in the requirement for a different kind of sensitivity calculation. It is introduced here. One may think of a stream of photons that generates a current of electrons, which are stored for an integration time. It is conceptually useful to separate the calculation into two steps. One is to calculate the photon irradiance on the aperture; the other is to calculate the number of electrons that result from that irradiance. Then the number of photons per second on the aperture of the optical system is reduced by the transmission factor τ_o for the optics, the fill factor η_{ff}, the quantum efficiency η, and the transfer efficiency η_t. Thus, the number of signal electrons that are generated in an integration time t_i in terms of the photon irradiance E_q on the aperture is given by

$$N_{se} = \tau_o \eta_{ff} \eta \eta_t E_q A_o t_i \, . \tag{6.31}$$

The photon number rate on the aperture is calculated in the same way as before, for point sources and extended sources. For a point source it is the source photon intensity times the solid angle the optics subtends at the source times the integration time.

$$E_{qo} = \frac{\tau_a I_q}{R^2} = \tau_a L_q \Omega_{so} = \tau_a L_q \frac{A_s}{R^2} \, . \tag{6.32}$$

For an extended source it is the source photon radiance times the *speed* solid angle of the optics (as described in Sec. 4.3.2).

$$E_{qo} = \tau_a L_q \Omega_{so} \, . \tag{6.33}$$

The noise arises from electron fluctuations that are caused by photon fluctuations in arrival rate and the leakage current.

$$N_{noise} = \sqrt{N_{se} + N_{leak}^2} \; t_i \, . \tag{6.34}$$

The number of electrons should be calculated rather than coulombs and amperes, but the latter should be used as a check for comparison with the rest of the electronics. The SNR must be adequate, and the well cannot be too full. Note that the noise in the photon stream is the square root of the mean number, so the mean square value is equal to the mean; that is why the first term in the square root is linear. The second term can be linear or squared, depending on whether the leakage current is expressed as an rms or mean square value.

Some examples are illustrative. Assume that a 3- to 5-µm system views a 300K blackbody scene. Imagine it has an F number of 3, a detector 100 µm on a side, atmospheric transmission of 1, optical transmission of 0.1, a quantum efficiency of 0.001, a fill factor of 0.25, a well capacity of 10^5, and an integration time of 1 msec. Substitution of these values into Eqs. (6.31) and (6.32), using the following values calculated with the programs shown in the Appendices,

$$L_{q3-5}(300) = 4.18 \times 10^{15} \quad \frac{1}{L_q} \frac{\partial L_q}{\partial T} = 0.04 \; , \tag{6.35}$$

shows that this will produce 10,000 electrons in a sample time, as follows.

$$N = \frac{\pi}{4} \frac{1 \times 0.135 \times 0.01 \times 0.25 \times 0.98 \times 0.001 \times 4.18 \times 10^{15} \times 10}{3.3^2} \tag{6.36}$$

$$N = 10,000 \; . \tag{6.37}$$

(I admit that I fudged the optical transmission to get this nice round answer). The noise from this is the square root or 100 electrons. The contrast for this system is 0.04, so that there are 400 electrons due to a change of one degree in the scene, and there is a noise of 100. This gives a SNR of 4 for a 1 K temperature difference in this spectral region. Equivalently it gives an NETD of 0.25. If this same system were operated in the 8- to 12-µm region, there would be about 50 times more photons and electrons, since there is more flux in the longer wavelength region. This factor can be calculated with the programs in the Appendices.

$$L_{q8-12} \simeq 50 L_{q3-5} \; . \tag{6.38}$$

Therefore, the well would be full all the time. That is not the way to do detection. Note that there are solutions for this. For a change, the solutions are to *reduce* the flux on the aperture or reduce the integration time.

6.3 The Emitted and Reflected Scene

An infrared sensor responds to the radiation that is incident upon its aperture. This incident radiation is always a combination of the radiation that is emitted from the scene and that which is emitted from the background as well as reflected by the scene to the sensor. We can represent the change in SNR that is generated when the sensor scans from one scene pixel to another by considering the total change in radiance, emitted and reflected. The total radiance is given by

$$L = \epsilon L^{BB} + \rho_b \epsilon L_b^{BB} \Omega_b \ , \tag{6.39}$$

where the first term is the radiance emitted from the scene and the second is the reflected background radiance. Note that the reflectivity is the bidirectional reflectivity. Since the background is not uniform, this should be represented as the sum of the different background radiances

$$L = \epsilon L^{BB} + \sum \rho_i \epsilon_i L_i^{BB} \Omega_i \ . \tag{6.40}$$

The change in SNR as the system scans from pixel to pixel is a result of the change in the temperature between the pixels and the change in emissivity between them. Of course, as the emissivity changes, so does the reflectivity. Therefore the change in radiance from point to point on the scene is given by differentiating the expression given in Eq. (6.38).

$$dL = \epsilon \frac{\partial L^{BB}}{\partial T} \, dT + L^{BB} \, d\epsilon + \sum \epsilon_i L_i^{BB} \Omega_i \, d\rho_i \ . \tag{6.41}$$

Recall that the reflectivity is related directly to the emissivity only for specular or Lambertian reflectors. The relative change can be written as

$$\frac{dL}{L} = \frac{1}{L^{BB}} \frac{\partial L^{BB}}{\partial T} dT + \frac{d\epsilon}{\epsilon} + \sum \frac{\epsilon_i L_i^{BB} \Omega_i \, d\rho_i}{\epsilon L^{BB}} \ . \tag{6.42}$$

Consider the first two terms that represent the change in emitted radiation. The first term is related directly to the NETD of the system, as it is the change in radiance due to a change in temperature. It can be evaluated for three important spectral bands, the total band, the 3- to 5-μm band and the 8- to 12-μm band. The first is easy. It is

$$\frac{dL}{L} = \frac{1}{L^{BB}} \frac{\partial L^{BB}}{\partial T} dT = 4d\frac{T}{T} \simeq \frac{4}{300} dT \simeq 0.01 dT \ . \tag{6.43}$$

For the 3- to 5-μm and 8- to 12-μm bands, the programs must be used for the evaluation. The results are 0.04 and 0.03. As will be shown in the next chapter, the emissivity throughout the infrared ranges from about 0.5 to 0.9, and this range does not change the rule of thumb appreciably. The 1%–1 K equivalence is nice. Therefore one can conclude that a 1 K change in temperature is about as effective in generating a change in radiance at the sensor as a 1% change in emissivity—and *vice versa*. The system NETD is not the whole story of how a system performs!

The reflected term provides some interesting information as well. Assume that the reflectivity is really Lambertian, so that the bidirectional reflectivity is ρ_h / π. Further, the hemispherical reflectivity is the one's complement of the emissivity. Then the change in radiance becomes

$$\frac{dL}{L} = \frac{1}{L^{BB}} \frac{\partial L^{BB}}{\partial T} \, dT + \left[1 - \sum \frac{\epsilon_i L_i^{BB} \Omega_i}{\pi L^{BB}} \right] \frac{d\epsilon}{\epsilon} \; . \qquad (6.44)$$

The bracketed term goes to zero when the entire hemisphere (with a projected solid angle of π) has blackbody radiation at the same temperature of the scene. This is washout.

7 ATMOSPHERIC TRANSMISSION AND RADIATION

Atmospheric transmission and emission are very important in the infrared region of the spectrum. In the visible region the atmosphere is essentially nonabsorbing, so the lack of transmission is due almost entirely to scattering. In the infrared there are regions in which there is almost total absorption and others in which the absorption is quite low. Scattering is minimal, since the wavelengths are so much longer and Rayleigh scattering depends on the inverse fourth power of the wavelength.

Atmospheric emissivity is essentially the ones complement of the atmospheric transmissivity. The transmission, or absorption, is a function of the composition of the atmosphere and the amount of absorption of the various important species.

7.1 Overview

Figure 7-1 shows a set of curves relating to the absorption of the different atmospheric species; it shows what is called a *solar spectrum*. Since the sun is essentially a blackbody in the infrared, the structure in the curves relates to the absorption of the various species. The infrared absorptions are due to molecular motions—rotations and vibrations. Atomic nitrogen and oxygen are not active in this region; they do not absorb. The absorbers are all listed in order from the top. They have the various absorption regions indicated and determined by laboratory measurements of so-called synthetic atmospheres. We can see that the significant regions are from 2 to 2.7 µm, at which point water vapor absorbs. Then, from 3 to 5 µm there is a window with a dip in the middle due to carbon dioxide. The big window from 8 to 12 µm is bounded on the short wave side by water vapor and on the long wave side by a combination of water vapor and carbon dioxide. The dip at about 10 µm is due to stratospheric ozone; it is not present in terrestrial viewing situations.

7.2 High-Resolution Spectra

For most infrared sensing applications, the low-resolution spectra are sufficient as guidelines, but the situation is much more complicated than is indicated by the previous figure. Figure 7-2 shows a small section of the atmospheric transmission curve with a resolution high enough to discriminate among the lines. After all, every one of these little absorption dips is due to a particular mode of motion of one of the constituent molecules, or an overtone!

7.3 Calculational Methods

There are two ways to calculate that make sense today. As a result of some phenomenal insight, the government saw fit to start, almost twenty years ago, a program to help us with these calculations. It is now possible to buy for about $60 a program called LOWTRAN 6 (I think you can get LOWTRAN 7 now) from the National Climatic Center, Federal Building, Asheville, North Carolina. It is a somewhat empirical program that is good enough for all but laserlike applications. The current version is on tape, in

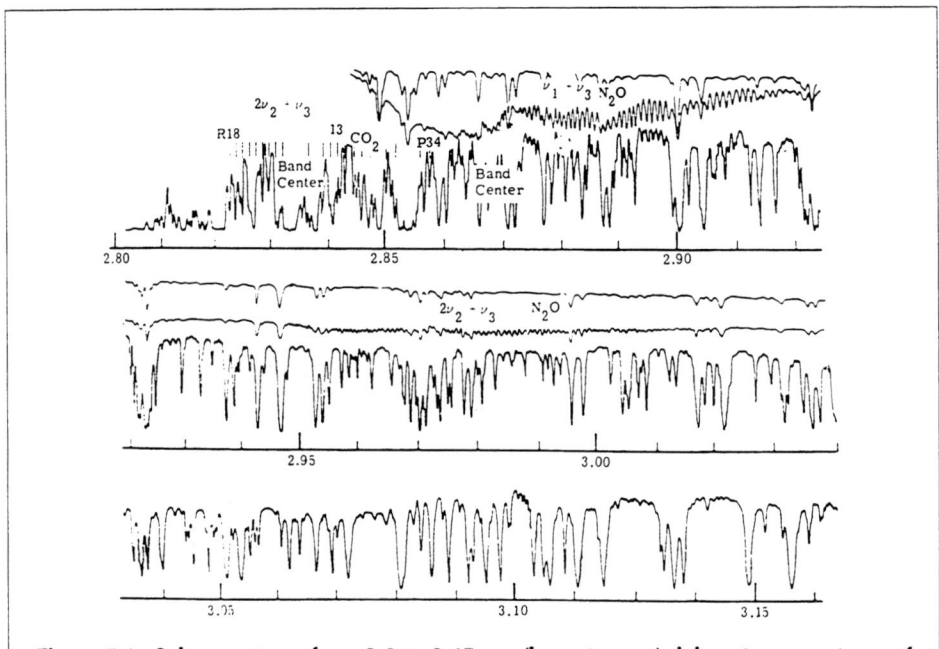

Figure 7-1. Solar spectrum from 2.8 to 3.15 μm (lowest curve); laboratory spectrum of H_2O (top curve); and laboratory spectrum of N_2O (middle curve).

FORTRAN, and can be run on most large systems. We have run ours on a VAX. Its major drawback is that it is card oriented. There are two solutions for this. One is to write a program that interprets the input and fills out the cards. Another is to spend about $500 with ONTAR[1] for PCTRAN. It is also a user-friendly system that fits on a PC. More recently a program using a better resolution, MODTRAN, that runs under Windows is also available; it is called PCModWin.

For laser lines, the corresponding program is FASCODE. This should also be used when the application requires viewing gaseous emission lines like CO_2 and H_2O through a similar atmosphere, like the air. In that event one has what I call the *comb* effect. There is a collection of lines from the source, each of which has been broadened because of the relatively high temperature. Each of these lines will be attenuated, mainly in the line center (a process called reversal). This is a high-resolution problem, and it must be dealt with by the use of a higher resolution program than LOWTRAN. I have had no direct experience with it, but the tape version can be obtained from the same source, and ONTAR has PC version for quite a bit more.

A typical run of a LOWTRAN 6/7 program takes about one minute to run on a 386 machine, depending on the length of the spectrum and the resolution. It obviously runs much faster on a 486DX100 or Pentium. It costs about $1.00 per run on a VAX.

---

[1] ONTAR Corporation, 9 Village Way, North Andover, MA 01845.

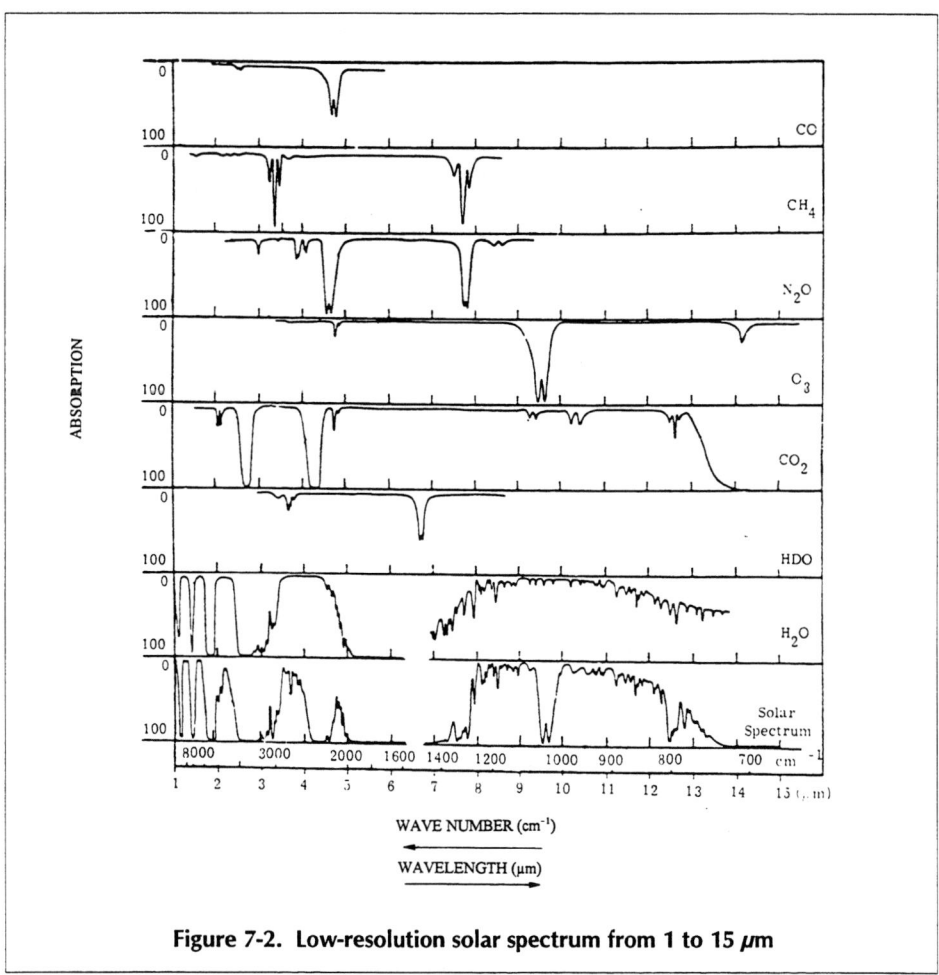

Figure 7-2. Low-resolution solar spectrum from 1 to 15 μm

runs much faster on a 486DX100 or Pentium. It costs about $1.00 per run on a VAX. (It doesn't cost anything on a PC). The spectra have been calculated at a resolution of 20 cm^{-1} but can be interpolated to 5 cm^{-1}. Although they can be run on a wavelength basis, they still do everything on a constant wave-number interval.

There are some papers on the use and comparisons of this code with experimental data, and the reader is referred to them.

7.4 Procedures

There are different options in running the ONTAR program, called PCTRAN. The one I prefer is to get into their program and type LOWIN. This is a program that inputs all the required data on a page-by-page basis. The beginning of the LOWIN program provides a listing on the monitor of the files that have already been run. You can use the arrow keys to choose one, or start a new program.

The first page requests things like the atmospheric model—1962 Standard, Maritime, Urban, user-defined, etc.; whether the profile should be suppressed; whether rain should be included; and the model for scattering. It also requests whether one wants just transmittance or radiance or both. I usually request radiance, because I get both radiance and transmittance, and I suppress the profile so I do not get a listing of all the atmospheric constituents every few meters! At the end, you are asked whether to change the name of the file. Do as you like. Only the input information, not the transmittances and radiances, are saved.

The second page, which is opened by the PAGE DOWN key on the keyboard, asks for geometry. This can be: (a) horizontal path, (b) slant path, or (c) slant path to space. These then require different sorts of inputs—(a) height and range, (b) first and second heights and range, and (c) height and zenith angle. The spectral range and resolution are also requested. Note that the second wave number of the range must be larger than the first.

The third page asks for plotting information. I have not used this much, as I almost always transfer the information to a QuickBASIC file and plot from there.

When LOWIN is done, the program LOWTRAN will calculate all the information you ever wanted. The program takes about one minute; just sit and watch. When it is almost done it creates a File7 file. The output data are in two places: the file you named with the extension .FL7 and in a file called LOWOUT. The latter has all of the output data plus all of the input data. The .FL7 file contains only the output (with some geometry and constituent data).

7.5 MODTRAN

This latest version (May 1995) has been written to work with Windows. The minimum requirements on your computer are a 386, 4 Mb of RAM, 6 Mb of hard drive, and a coprocessor, although more capacity will surely make things go more smoothly. ONTAR recommends at least 5 Mb for file space, results, and the like. I have not yet had an opportunity to "wring out" this new version, so I cannot provide additional description.

8 EVALUATION OF THE IR SCENE

The input signal-to-noise ratio (SNR) is theoretically a ratio of blackbody radiations from the scene. It is a good starting place, so we will evaluate it; in many cases its value is not far from a very carefully evaluated actual scene. Then we will look at some data about the real scene, although this is a highly variable and highly statistical set of data.

8.1 Evaluation of the Scene SNR

The idealized SNR for a temperature sensor was shown to be

$$\frac{1}{\text{NETD}} = \frac{\partial \text{SNR}}{\partial T} = \frac{0.74}{\sqrt{B}} \left[\bar{\frac{1}{\lambda}} \frac{\int \frac{\partial L_q}{\partial T}\, \mathrm{d}\lambda}{\sqrt{\int L_q\, \mathrm{d}\lambda}} \right]. \tag{8.1}$$

The bracketed term represents the wavelength that was used to calculate the diffraction limit, times the change in signal in the spectral passband for a one-degree change in temperature from pixel to pixel in the scene, divided by the square root of the background flux in the passband, the radiation noise. The wavelength may be the maximum or the mid wavelength or the average wavelength weighted by the spectral responsivity of the system. The radiation terms are given in radiances, for ease of comparison, and we have put them in the form of quantum fluxes because photon detection was assumed. Photon detection is the most sensitive, and we will probably need all we can get. The bracket is easily evaluated by the computer programs provided in the Appendices. The result is shown in Fig. 8-1. The values for several different

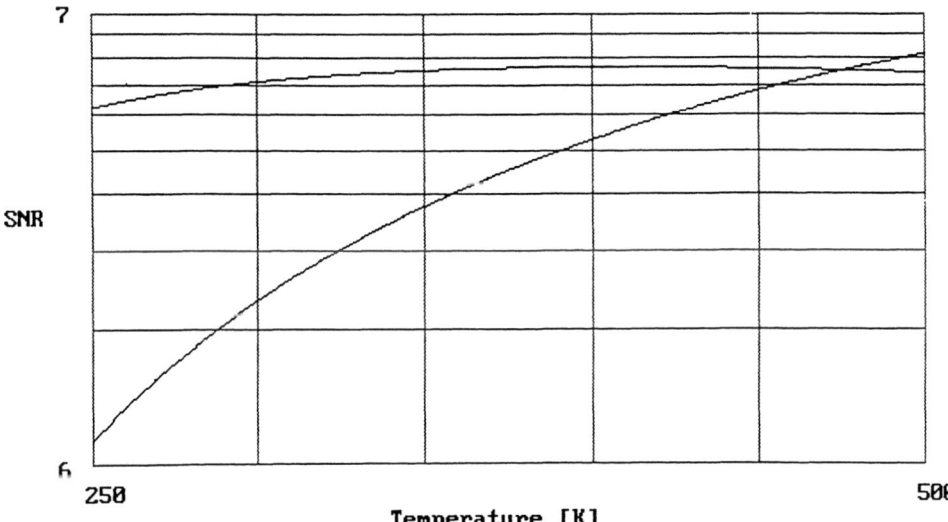

Figure 8-1. Radiation signal-to-noise ratio

temperatures are shown for two spectral regions, the 3- to 5-μm and 8- to 12-μm bands. These are the two most important transmission bands of the atmosphere, as we saw in the previous chapter. Notice that the 8- to 12-μm band has almost ten times the input SNR as the 3- to 5-μm band. This arises because there is a greater change in the flux with respect to temperature right around 8 μm for a 300 K blackbody. The overall SNR will have an even greater difference because the wavelength is in the numerator, and the ratio is almost two, as these two curves show. Other things being equal, we should use the longer band when doing thermal imaging. But other things are not equal. An article in *Applied Optics*[1] provides a more detailed evaluation that includes the effects of the atmosphere.

8.2 The Nature of the Infrared Scene

It is impossible to describe in detail all the different aspects of the world as seen in the infrared part of the spectrum. Only some generalizations can be made. First, beyond about 5 μm, almost everything is approximately black. This includes snow, concrete, and white paint as well as the dirt and asphalt and other things we think of as black. Black in the visible is not necessarily black in the infrared! And white in the visible is not always white in the infrared, either.

8.2.1 A Look Down

The Earth as seen from satellite altitude, shown in Figure 8-2, consists of many clouds and a more or less black Earth at about 300 K (depending on season and location). The curve was calculated for a portion of the Earth that is not sunlit, between 30 and 40 degrees North Latitude and for three different γ' angles, where γ' is the angle the line of sight from the Earth intersection point makes with the radius of the Earth. The clouds are largely black and take on a temperature appropriate for their altitude. The curves are for a clear-sky view of the earth. They can be bounded generally by two blackbody curves, one at the temperature of the earth, and the other at about 250 K, representing the temperature of the effective top of the atmosphere. One sees with his infrared eyes, as he looks down, the radiation from the top of the atmosphere, where the atmosphere is opaque, and the radiation from the earth, where the atmosphere is transparent.

The general nature of the scene in the transparent regions is that the radiance increases with both sun angle and with decreasing latitude. Neither of these is a direct effect of sunlight. Both effects result from the increased heating by the sun. We all know that, as the sun gets higher in the sky, the day gets warmer, and after noon (or at least maximum sun angle), the day begins to cool. We also know that northern latitudes are colder than southern ones. This is a direct consequence of the projection affect of the globe. The decreased input flux from the sun results in a cooler ground, resulting in less emitted radiation. Again, neither of these two effects is related to reflected sunlight.

---

[1] A. Findlay and D. R. Cutten, "Comparison of performance of 8-12 and 3-5 μm infrared systems," *Applied Optics* **28**, 4961 (1989)

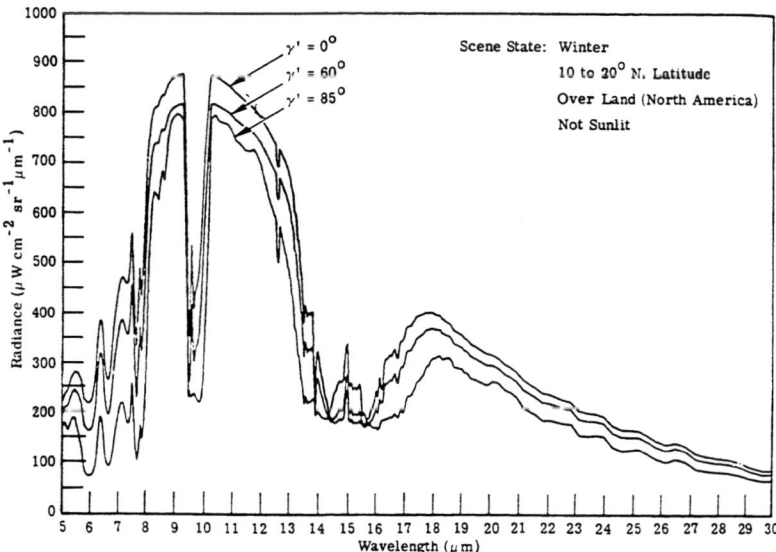

Figure 8-2. A look down

Although clouds are not shown here, we may assume that they are blackbodies at their equilibrium temperature, about the air temperature at their altitude.

8.2.2 The Infrared Sun

The sun is not our undoing. It is our main source of light. Its spectral distribution is shown in Fig. 8-3. The sky is an important radiator in the infrared. The amount of radiation is a function of the spectral region, the local content of the atmosphere, and the

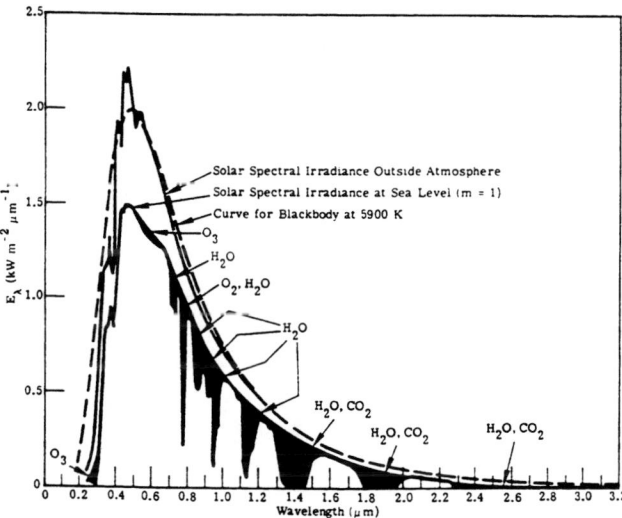

Figure 8-3. The infrared sun

zenith or nadir angle of view. More of this when we come to atmospheric transmission. The solar spectral irradiance curve is extrapolated to be above the atmosphere. In fact, the structure you see in the curve represents the different atmospheric absorptions. The Stephan-Boltzmann law for total radiation indicates that

$$M = \sigma T^4 . \tag{8.2}$$

The radiance of the sun is given by

$$L = \frac{M}{\pi} = \frac{\sigma T^4}{\pi} = \frac{5.67 \times 10^{-8} \times 5900^4}{3.14159} = 21869620 . \tag{8.3}$$

The irradiance at the top of the earth's atmosphere is given by

$$E = L\Omega = \frac{\pi}{4}\left(\frac{34}{60}\frac{\pi}{180}\right)^2 L = 7.68 \times 10^{-5} = 1386 Wm^{-2} . \tag{8.4}$$

This compares very well with the measured solar constant (the total flux density at the top of the atmosphere from the zenith sun) of 1380 Wm^{-2}.

It can be shown without much trouble that the spectral solar flux incident on the earth is equal to the flux emitted from the earth at about 4.5 μm and is correspondingly less at longer wavelengths. The process is to set the blackbody spectral solar irradiance equal to the blackbody spectral solar emittance of the earth. The sun is a blackbody of 5900 K and the earth one of 300 K, but the sunlight is attenuated by its distance. Thus the equality is that the flux density at the top of the atmosphere from the sun is equal to the flux density at that point from the emission from the earth. The equality can be set up as

$$L^{BB}(T_{sun})\Omega_{sun} = L^{BB}(T_{earth})\Omega_{earth} . \tag{8.5}$$

$$\frac{c_1\Omega_{sun}}{\lambda^5(e^{c_2/\lambda T_{sun}} - 1)} = \frac{c_1\Omega_{earth}}{\lambda^5(e^{c_2/\lambda T_{earth}} - 1)} . \tag{8.6}$$

Much of this can be eliminated so that it reduces to

$$\frac{\Omega_{sun}}{(e^{c_2/\lambda T_{sun}} - 1)} = \frac{\Omega_{earth}}{(e^{c_2/\lambda T_{earth}} - 1)} . \tag{8.7}$$

This can be rearranged and the ratio of solid angles calculated; that of the sun was evaluated above as 5.68×10^5, while that of the earth, right next to the point of evaluation is π. In that case,

$$\frac{(e^{c2/\lambda T_{sun}}-1)}{(e^{c2/\lambda T_{earth}}-1)} = \frac{\Omega_{sun}}{\Omega_{earth}} = 2.445\times10^{-5} \ . \tag{8.8}$$

If the Wien approximation can be used, then the ones are dropped and

$$\frac{(e^{c2/\lambda T_{sun}})}{(e^{c2/\lambda T_{earth}})} = e^{\frac{c_2}{\lambda}\left(\frac{1}{T_{sun}}-\frac{1}{T_{earth}}\right)} = \frac{\Omega_{sun}}{\Omega_{earth}} = 2.445\times10^{-5} \ . \tag{8.9}$$

The result of this calculation is that $\lambda = 4.29$. The real answer, using the Planck equation, is 4.63 μm.

8.2.3 The Infrared Sky

Figure 8-4, taken from *The Infrared Handbook,* shows the radiance of the clear zenith sky as a function of wavelength. It is not really less than zero at 11 μm, but it is low. The radiance of the sky is very low where the atmosphere is transparent, and it increases in the regions of absorption. Figure 8-5, *ibid,* shows the radiance of the sky near the horizon. There is almost no horizon; the radiance is the same as the earth's in the whole spectral region. This is a manifestation of the very long atmospheric path at the horizon!

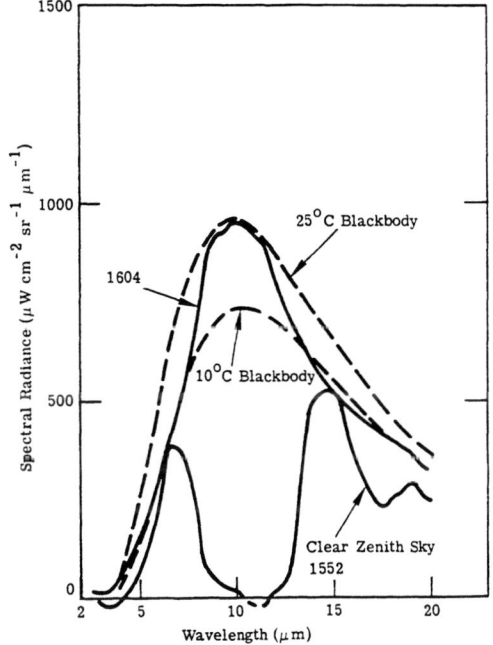

Figure 8-4. The infrared sky

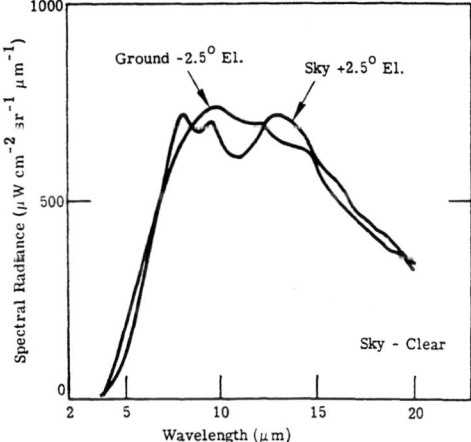

Figure 8-5. The infrared horizon

8.2.4 The Terrestrial Scene

Figure 8-6, from *The Infrared Handbook,* shows reflectances. Since the materials are opaque, the emissivity is the ones complement of these values (assuming that the measurements were hemispherical reflectivity). Soils have the general characteristic of a 50% reflectance (and 50% emissivity) from the visible to almost 3 μm. They recoup the reflectance (almost) from 3 to 5 μm and then drop to a consistent and low reflectivity value of about 10% (emissivity of 90%). There is a little structure in the 10 μm region. The increased emissivity around 3 μm is due to the absorption of water. Beach sand in the visible has the highest reflectivity, and the red dirt of Georgia is not far behind. They are almost identical beyond 5 μm. There is a noticeable rise in reflectivity in the region from 8 to 10 μm; this is a reststrahlen (residual ray) region. Each of the different minerals has a slightly different reflection spectrum in this region of anomalous dispersion. Good spectroscopy can analyze the mineral content of the soils!

Road surfaces are also shown. They look a lot like dirt, when the differences in scale of the plots are considered. This is the kind of information that generates the contrasts we get in infrared night driving systems and the information we obtain from the infrared portion of the remote-sensing satellites.

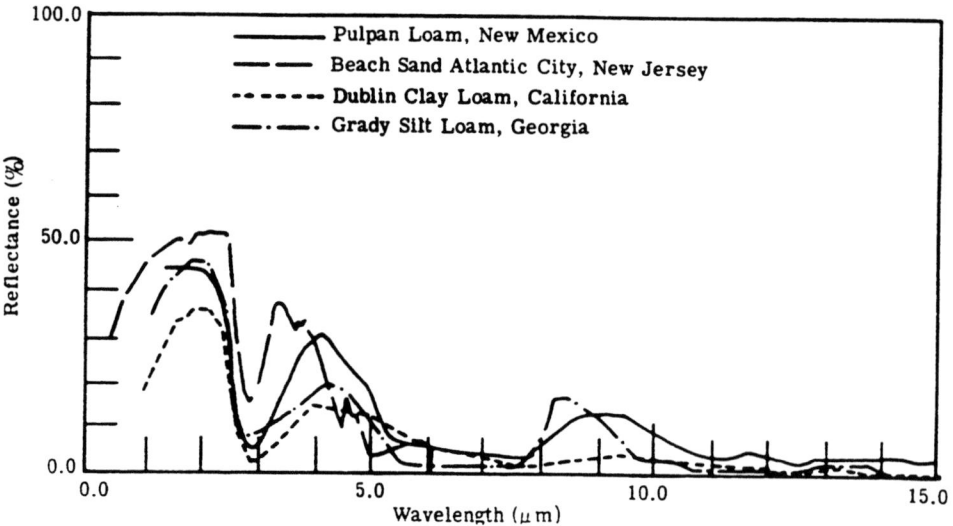

Figure 8-6. Terrestrial data

8.2.5 Diurnal Variations

The sun heats the earth by day, and the (Arizona) sky lets it cool by night. It is not unusual to expect, therefore, that the radiation in the infrared is different by day and by night. These variations are shown in Fig. 8-7. The top part of the figure shows that in the near infrared different objects "emit" by reflection. The various materials—grass, snow, concrete, red bricks—all have reflectivities that are somewhat familiar to us. Their relative temperatures are due to these reflectivities and the manner and amount of solar radiation they reflect. This condition is reasonably familiar to our senses from the visible portion of the spectrum. But at a wavelength of about 3.5 µm the situation changes. All have increasing radiance with wavelength, rather than decreasing. They are riding on the blackbody spectrum of the earth, at about 300 K, which peaks at about 10 µm, the spectral region of these data. They are all very similar because they all have about the same emissivities, as we have already seen.

Therefore, the emission is almost blackbody in nature and close to 300 K, but directly related to ambient. The ambient swings from about 5 C to about 20 C from day to night. The swing is more pronounced in arid areas, like Tucson, where the atmosphere is particularly transparent. Some things cool faster than others; it is a question of their heat capacity and mass. Therefore we can see the radiation plots of various materials in the infrared scene cross. They all come fairly close together shortly after sunset and after sunrise. These are the crossover and washout times.

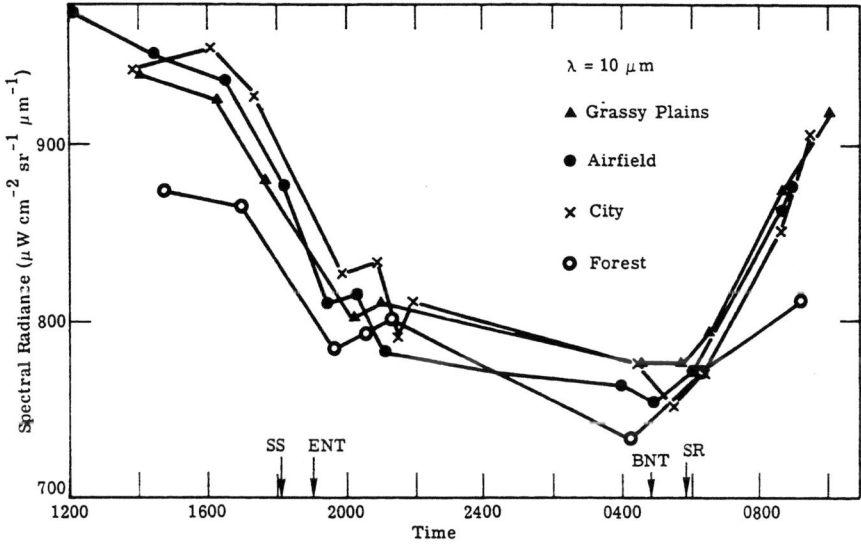

Figure 8-7. Diurnal variations. SS = sunset; SR = sunrise; ENT = end of nautical twilight; BNT = beginning of nautical twilight.

8.2.6 People

The emissivity of people, shown in Figure 8-8, is important in that we want to detect them for night-driving purposes and we want to diagnose them for health reasons. The skin of everyone, regardless of race, creed, or color, is very black from 5 μm on. It is somewhat less black in the 3- to 5-μm region, and of course it varies in the visible. The emissivity of the human skin is hard to measure because the temperature is hard to measure. It is surely a function of the amount of perspiration.

These are data on the emissivity of the skin. The temperature of a human being is known to be 98.6° F, which translates to 37° C and 310 K. This makes the nude human a very good target in the terrestrial environment. Of course it is unusual for people to run around in the nude, but their faces and hands almost always show. In addition, most people are in close contact with their clothing. Cloth has an emissivity almost as high as the human skin. A nice coincidence is that when people are all bundled up, they are that way because it is cold outside. The contrast is maintained reasonably well. These features of the clothes and the human have also given rise to the myth that one can see through clothes with infrared. One sees the heat pattern caused by contact with clothes, unless the clothes are polyethylene, as will be seen in Chap. 10.

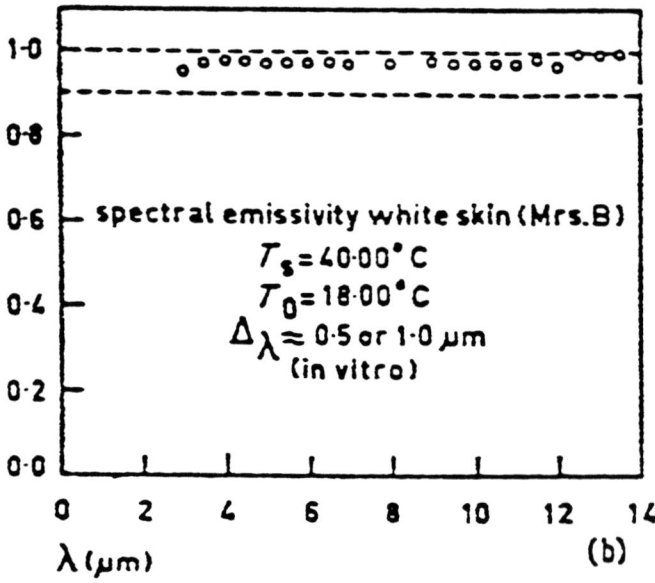

Figure 8-8. People

8.2.7 Paints

Some paints are shown in Figure 8-9. They are mostly black as a result of the binders and leads in them. The metallic paints, reasonably enough, are not black, and thin layers on metals may not be. A good, thick white paint is black in the infrared, and a good, thick black paint is black in the infrared... as well as a red one, a purple one and so on. For those involved with military applications, it is good to note that just about all camouflage paints look alike.

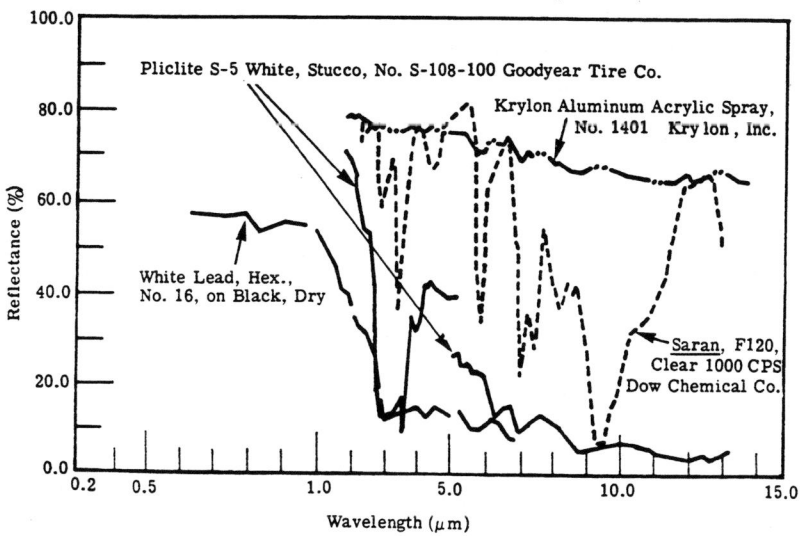

Figure 8-9. Paints

8.2.8 Some Corroborative Data

Investigators at Syracuse University took some spectral irradiance data back in the 1950s. Figure 8-10 shows the results of scans in the visible and near infrared regions. They support the calculations we made earlier about the nature of radiation in this region.

The curves on the left represent reflected sunlight. Snow, which we know is white (at least when it is clean), has the highest set of values because it has the highest reflectivity. Then come the red brick wall, concrete (probably quite gray), and grass. It is hard to tell about colors, since this is all for wavelengths above 1 μm.

Starting at 3 μm the curves[2] all show a rise with increasing wavelength, characteristic of the rise of the 300 K blackbody curve. They bunch together because their emissivities are all much alike. Although the inflection point is not at 4.5 μm, it is close, and the difference is due to the affects of reflectivity and emissivity. An

---

[2] W. R. Frederickson, N. Ginsberg, and R. Paulson, *Infrared Spectral Emissivity of Terrain,* Final Report, Syracuse University Research Institute, Syracuse, NY.

interesting exercise is to use the programs in the Appendices and incorporate the emissivity values to see where the crossover point is.

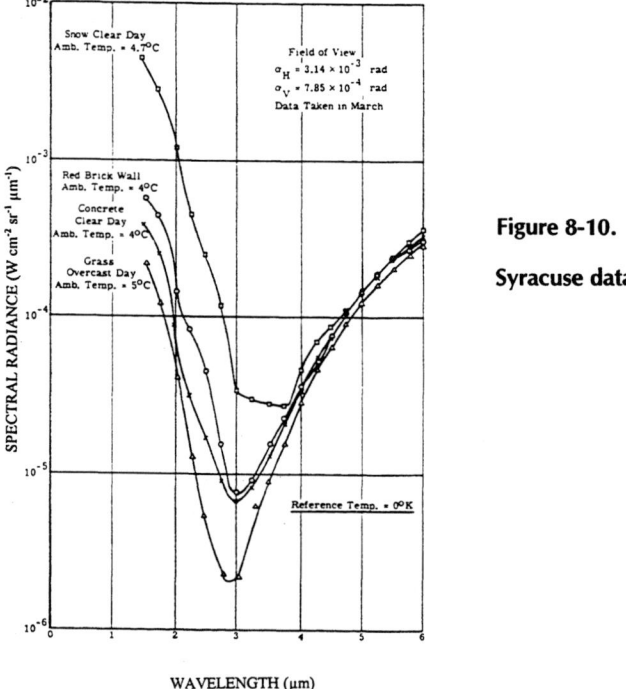

Figure 8-10.

Syracuse data

WAVELENGTH (μm)

8.2.9 Water

Notice in the strange curve of the absorption coefficient of water, Fig. 8-11, the enormous peak in the wavelength region beyond about 2 μm. The curve is a plot of the absorption coefficient of water in reciprocal centimeters as a function of wave number, although wavelength is cited on the abscissa, running from right to left.

The absorption coefficient, which will be called β here, is the proportionality constant in Beer's law. The transmission is

$$\tau = e^{-\beta x} . \qquad (8.10)$$

Neglecting reflectivity, the absorptivity α is 1 minus this value, and for small values is approximately βx,

$$\alpha = 1 - \tau = 1 - e^{-\beta x} \approx 1 - (1 - \beta x) = \beta x . \qquad (8.11)$$

The coefficient on the chart is given in reciprocal centimeters, as it usually is. If you divide it by 10 to get reciprocal millimeters, you will find that for an absorption coefficient of 10,000 cm^{-1} the absorptivity (equal to the emissivity) is 1000 mm^{-1}. One millimeter of water then absorbs all the radiation to the number of digits my calculator can handle. The transmission of 0.1 mm is 3.7×10^{-44}; that of .01 mm, 10 μm, is 4.5×10^{-5}; and 1 μm transmits 36% (absorbs 64%). Three micrometers then absorb 95%. Almost all of the radiation is absorbed in a few micrometers of water. This explains much of the observed data on people, paints, and things. Anything that has any appreciable amount of water in it has a high absorptivity and a high emissivity.

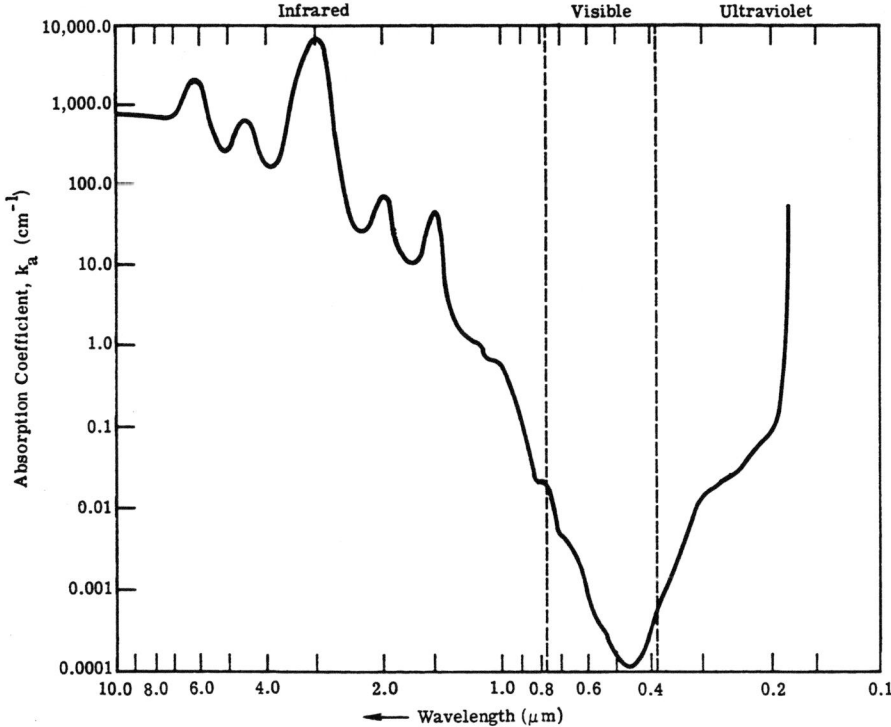

Figure 8-11. Water

This has also been one of the downfalls of infrared diagnosis. In the hunt for the generic Red October, an airborne infrared system can only "see" any changes that are brought to the surface. The submarine usually exhausts warm effluent, but it turns over lower, cooler water to the top. Sometimes the wake is warmer and sometimes it is colder, but sometimes it is just right or wrong. The diagnosis of cancer in people has some of the same characteristics. Only the diffusion to the surface can be sensed, and sometimes this is misleading. An analysis of the effluent of TVA power plants was considered many years ago. Although the heated plume that flows into the lake or river may almost always be seen, it is extremely difficult to quantify.

9 BANDWIDTH AND SCAN STRATEGIES

The electrical bandwidth of the infrared system is a determinant of its performance. The larger the bandwidth, the more the noise and the poorer the performance, by the square root of the bandwidth. The information bandwidth is determined by the rate at which information must be transferred, and this determines the noise.

This chapter provides the methods by which the calculation of the information bandwidth can be made and its relation to the effective noise bandwidth.

9.1 Information Bandwidth

There is a very encompassing theorem of Claude Shannon that says simply that the minimum frequency needed for complete transfer of information is the reciprocal of twice the dwell time or the shortest time interval involved with measuring the data. Every infrared system has a number of resolution elements or pixels in its field of view. This is true whether it is an imaging device of television bandwidth, with about 500 pixels in a line and 500 lines in a frame with 30 frames per second, or whether it is a tracker or surveillance set that covers 360 × 10 degrees or an entire hemisphere. In a few degenerate cases there is only one pixel in the entire field, but the following discussion still applies.

Figure 9-1 shows an elementary frame. It consists of a rectangular collection of

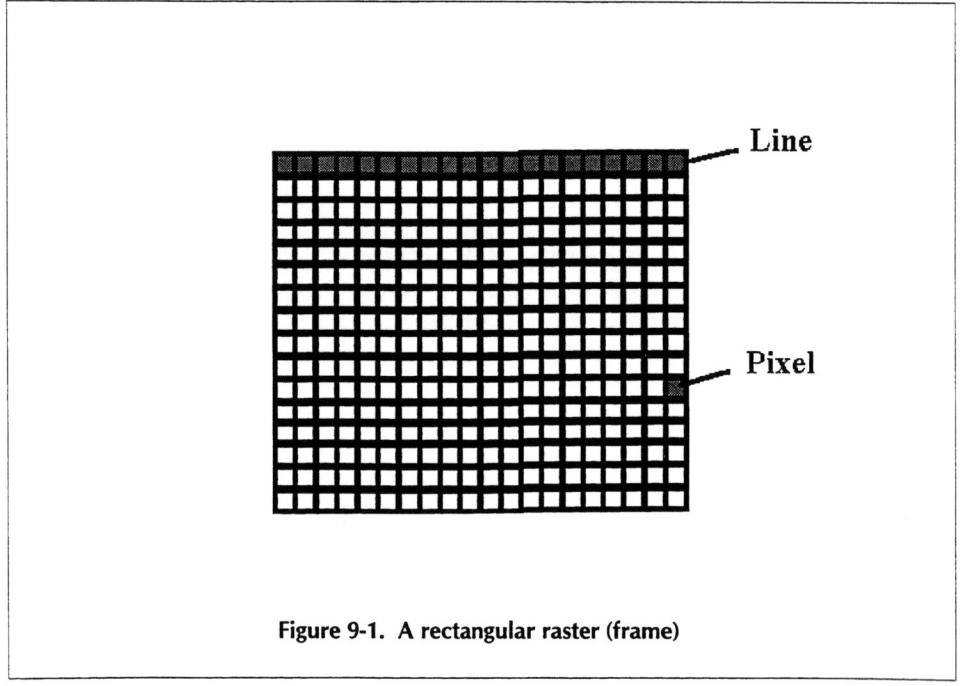

Figure 9-1. A rectangular raster (frame)

pixels, arranged in horizontal lines. As stated above, the required information bandwidth is given by

$$B = \frac{1}{2t_d} ,$$ (9.1)

where B is the bandwidth and t_d is the dwell time. The ideal dwell time is given by

$$t_d = \frac{t_f}{N} ,$$ (9.2)

where t_f is the frame time and N is the number of pixels in the frame. The number of pixels in the frame N can be determined in several ways. It is the solid angle of the frame divided by the solid angle of a pixel. It is the number of lines times the number of pixels in a line. It is the number of horizontal pixels n_h (in a row) times the number of vertical pixels n_v (in a column), and it is the square of the number of lines times the aspect ratio. Then the bandwidth is

$$B = \frac{n_h n_v}{2t_f} .$$ (9.3)

This is the expression for the idealized bandwidth, one in which there is no dead time, no time for flyback, constant scan velocity, a system that almost never exists. (I can think of just one example.) When the system is not ideal, a scan efficiency η_{sc} takes it into account (as shown below). The somewhat circular definition of the scan efficiency is the ratio of the real dwell time to the ideal dwell time. (The real dwell time will always be shorter.) It can be evaluated for specific systems based on their individual characteristics. When more than one detector is used, the dwell time can be lengthened and the bandwidth decreased. There are several ways to do this that will be discussed in the next few sections. At any rate, the bandwidth is decreased (or effectively decreased) by the number of detectors m that are used in the system. Thus the equation for bandwidth is

$$B = \frac{n_h n_v}{2m\eta_{sc}t_f} .$$ (9.4)

This can be applied to a more or less standard television system that has 500 lines and a 3 to 4 aspect ratio, one detector, and virtually 100% scan efficiency (since it is an electron beam scanner). Then

$$B = \frac{3}{4} \frac{500 \times 500}{2/30} = 2812500 \; . \tag{9.5}$$

TV requires a bandwidth greater than 3 MHz to account for all the pixels and their gray-scale levels. If this were an infrared scanner with a single detector, the bandwidth would be the same (assuming 100% scan efficiency), and the dwell time on any pixel would be about 0.1 μs.

The concepts can be applied to a so-called strip mapper, a system carried in a plane or satellite that sweeps out line after line transverse to its forward motion, while it generates a strip of unending, or at least indefinite length. The transverse sweeps or lines must be contiguous at the nadir, right under the vehicle. This line width is $h\alpha$, where h is the vehicle height and α is the angular size of a pixel. The time allowed for contiguous lines then is this distance divided by the velocity v.

$$t_l = \frac{h\alpha}{v} \; . \tag{9.6}$$

The dwell time is the line time divided by the number of pixels in a line, which is the total scanned angle divided by the angular pixel size.

$$t_d = \frac{t_l}{n_l} = \frac{t_l}{\Theta/\alpha} = \frac{h}{v} \frac{\alpha^2}{\Theta} \; . \tag{9.7}$$

This does not include the number of detectors or the scan efficiency. Solving for the bandwidth, with these included, provides

$$B = \frac{v}{h} \frac{\Theta}{2\eta_{sc} m \alpha^2} \; . \tag{9.8}$$

The lower and faster a vehicle flies, the tougher the problem. The higher the resolution (the smaller the pixel subtense), the really tougher the problem.

9.2 Time Delay and Integration

It was tacitly assumed that the use of a number of detectors in a linear array could reduce the bandwidth by some sort of time sharing. One obvious way to do this is to align them perpendicular to the scan direction, as shown in Fig. 9-2. If there are as many detectors as there are lines, then the line time is equal to the frame time, and the dwell time is the frame time divided by the number of pixels in a line. This is reasonably intuitive. However, there is an alternate scheme.

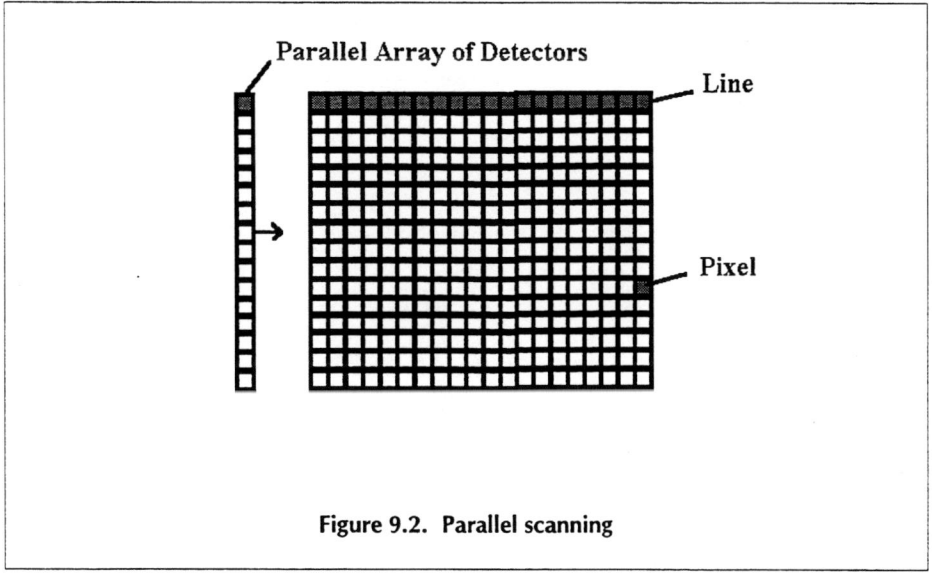

Figure 9.2. Parallel scanning

If the array, with far fewer detectors than pixels in a line, is oriented parallel to the scan direction, then as a line is scanned, each detector "sees" each pixel in sequence. This is illustrated in Fig. 9-3. If the signal input obtained by the first detector element in the array is stored for the time it takes for the next detector element to "see" the same pixel

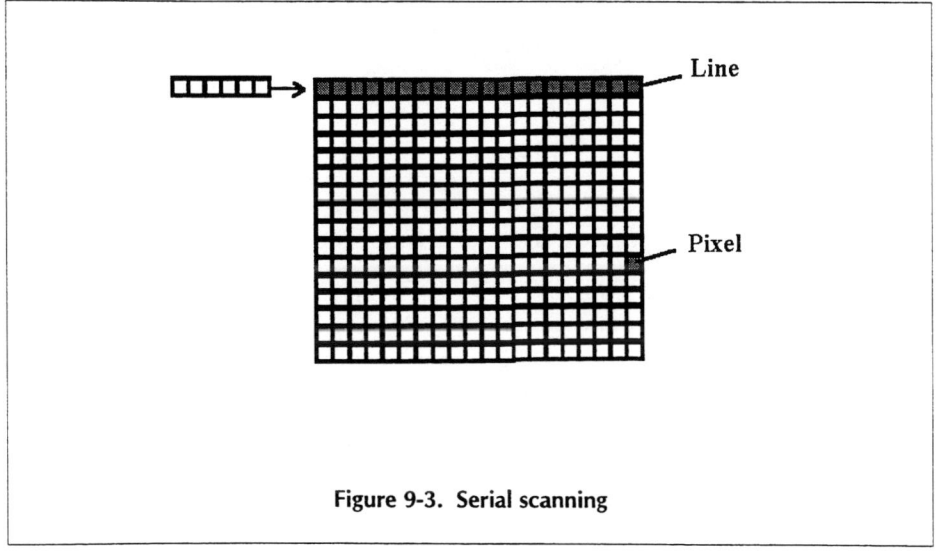

Figure 9-3. Serial scanning

and these are added, the signal from that pixel is increased (doubled). If this is repeated, for m elements, the signal will be increased m times. The noise will also be increased, but by the square root of m. Thus, the SNR is increased by the square root of m. This technique is called *time delay and integration,* although it is really time delay and addition. The abbreviation, logically enough, is TDI.

A TDI scanner must scan as fast as a single-detector scanner, but it has the same effect as reducing the bandwidth: it increases the SNR by the square root of m. Since every element senses the radiation from every pixel, any nonuniformity in the array is unimportant. The other advantage of TDI is the simplicity of dc restoration and calibration—since the array can be made to sense a calibration source at the end of each scan line.

The idea of time delay, if not integration, can also be applied to staggered arrays, as shown in Fig. 9-4. It can make the array easier to make by separating vertical lines, and it can allow the use of staggered arrays that would otherwise be impossible.

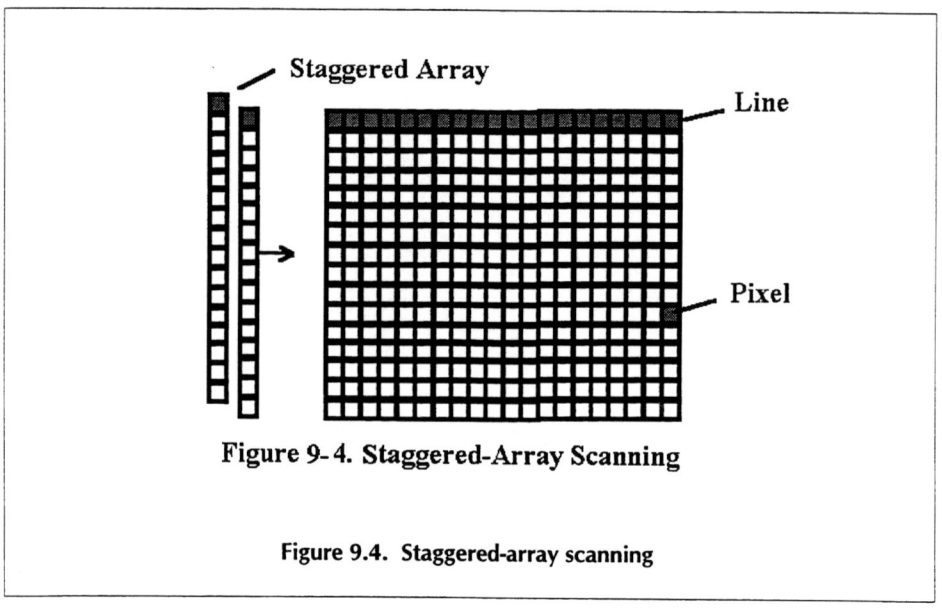

Figure 9- 4. Staggered-Array Scanning

Figure 9.4. Staggered-array scanning

The concepts of parallel scan and TDI, or serial scan, can be combined to produce a so-called hybrid scan. This uses a two-dimensional (staggered) array in a scanning mode, as shown in Fig. 9-5. The array is scanned across the field of view, using TDI in the rows. Such an array is usually 5 to 10 columns with many rows, and then another 5 to 10 columns with as many rows offset from the first by half a pitch. Therefore there is complete, overlapped coverage of the field.

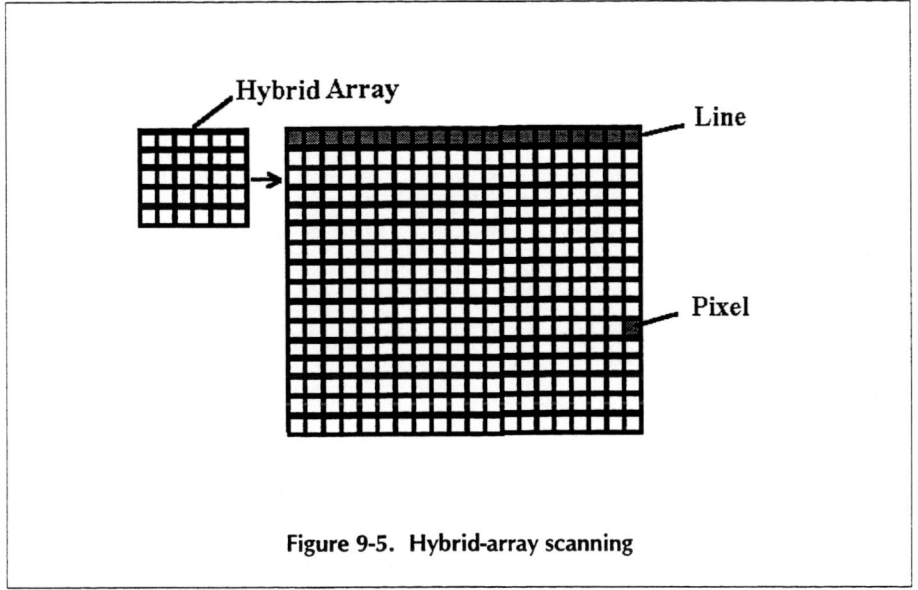

Figure 9-5. Hybrid-array scanning

9.3 Optimizing Scan Patterns

Optimization is most important for scan patterns that have high information content, such as television images. The slower the required frame rate and the fewer the pixels, and the more you can get away with, the less efficient you need to be. This discussion involves doing the best you can. A less demanding system can have simpler solutions.

The first consideration is that the dwell time must be no shorter than the responsive time constant of a detector. Recall that when the dwell time equals the response time, the signal only rises to 63% of the maximum. Two dwell times give 95% and three about 99%. It is obviously important to keep the scan efficiency as high as possible. It is not so obvious that the two ways to do this are to keep the dead time at a minimum and to keep the scan velocity constant. The first of these is accomplished by not using flyback and by specific design techniques with scanners, usually rotary, rather than oscillatory.

If a scanner has a variable velocity, for instance a resonant scanner, the bandwidth must be set according to the *minimum* dwell time. The ideal dwell time is the average dwell time over the field, but the minimum dwell time of a variable-speed scanner will be less than the ideal, and the bandwidth will be greater than ideal.

There are two good reasons for unidirectional scanning: spatial and temporal. Figure 9-6 shows the effects of scanning in two different directions on a temporal basis; Fig. 9-7 shows the spatial counterpart. As one scans from left to right across a rectangle, like a road or railroad track, the left-to-right scan peaks in a different place than the right-to-left scan. One is caused by the finite rise time; the other is caused by the convolution of the detector image across the object.

So the rules are: scan with constant velocity in the same direction with no dwell time. This is a wonderful challenge.

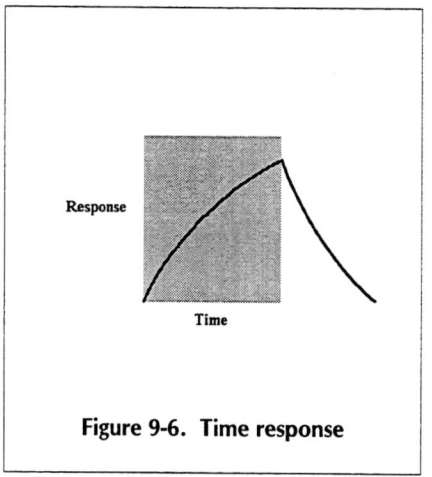

Figure 9-6. Time response

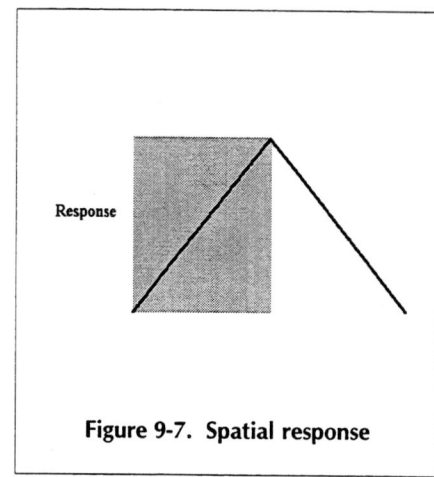

Figure 9-7. Spatial response

9.4 The Bow-Tie Effect

When an aircraft or satellite scans to obtain a strip map, it scans a detector or array image from side to side. (Of course, this should be unidirectional). The size and shape of the image of a square detector changes with field angle. This happens for two reasons: the line-of-sight distance increases, and the image must be projected to the ground. It is useful to assume that the ground is flat. The curvature of the Earth is marginal, and terrain features vary too much for any general analysis. The square pixel at the ground directly below the vehicle is $2h\tan(\alpha/2)$. For most purposes this can be approximated by αh, since the pixel angle will almost always be small. The size of the pixel on the ground at a field angle of $\Theta/2$ is $R\tan\alpha$ where R is the slant range, given by $h/\cot\Theta/2$. Therefore the pixel has a nearer side given by $h\tan\alpha/\cot\Theta/2$, while the outside side is $h\tan\alpha/\cot(\Theta/2+\alpha)$. This trapezoid is shown in Fig. 9-8. The entire lateral scan has the general shape of a bow tie.

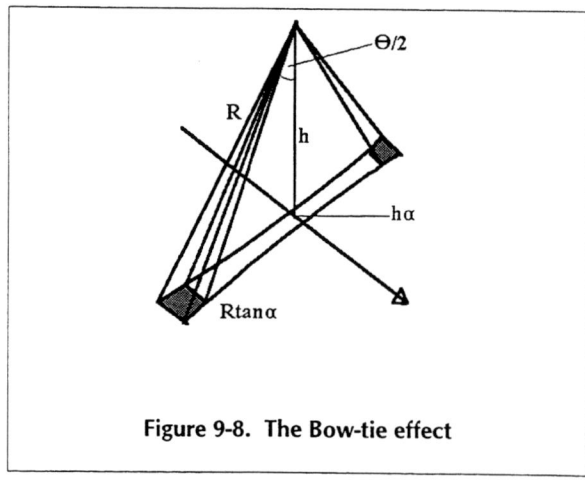

Figure 9-8. The Bow-tie effect

9.5 The Palmer Scan

One way to avoid the bow-tie effect is with a Palmer scan. This is a pattern that is generated by translating a circular scan, as is shown in Fig. 9-9. If the vehicle makes the translation, all that is necessary of the optical system is a circular scan. The pixel is always the same size and shape, since the slant range is always the same. However, for identical systems, the Palmer scan never has as good resolution as the simple transverse scan at nadir. For some applications, such as scanning hot-rolled strip steel, in which the object distance is relatively small, the depth of field is important. The Palmer scanner also provides a constant object distance so that the depth-of-field problem is ameliorated.

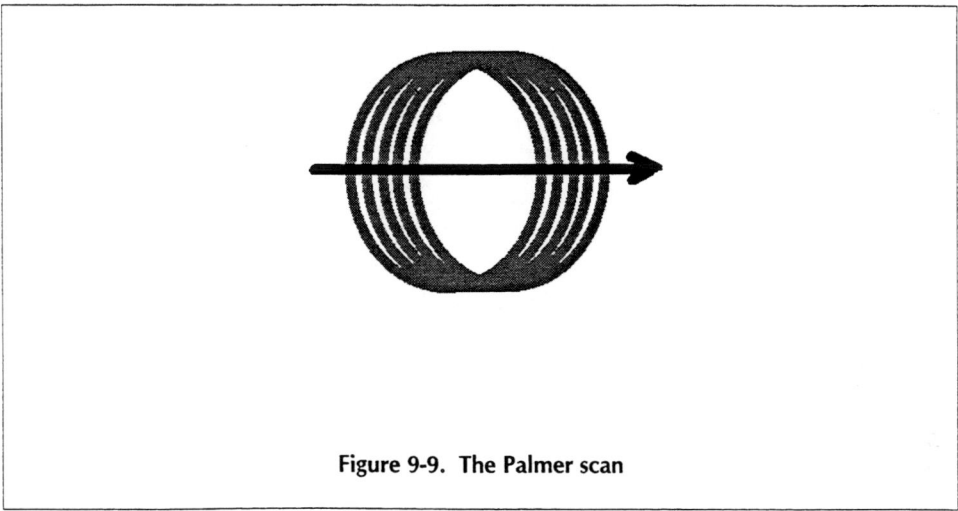

Figure 9-9. The Palmer scan

9.6 Effective Noise Bandwidth

All these considerations lead to the maximum frequency of the information bandwidth. The noise, however, has a (nonconstant) spectrum. It is not flat. For purposes of calculations, however, it is useful to accept the fiction that it is flat. So one must calculate the effective noise bandwidth and the effective noise. If B is the information bandwidth and $V(f)$ is the noise voltage spectrum, then the effective noise bandwidth is given by KB, where K is given by

$$K = \frac{\sqrt{\int V^2(f)df}}{B} .$$ (9.9)

This is a useful fiction for preliminary calculations, but as usual, all the details must be taken into account when the strokes get short.

10 OPTICAL MATERIALS

In the visible portion of the spectrum almost all lenses and windows are made of silicate glass. It is transparent, strong, hard, resists chemical attack, and is a wonderful material overall. Unfortunately, silicate glass transmits only to about 2 μm. Therefore many different materials must be used in different parts of the infrared spectrum.

The choice between the use of lenses and mirrors leans more toward mirrors in the infrared than in the visible. There are several reasons for this, which will be discussed. In addition, the choice criteria and some of the shorts and longs of each of these will be discussed. The reader is cautioned to use this information as a beginning guide and to review properties and performances with individual vendors.

10.1 Types of Materials

There exist infrared glasses, although they do not have any silica in them. Single crystals are used in some applications, even though their size is limited. Polycrystalline materials and even oligocrystals (just a few crystallites) are used, and there is controversy regarding the use of single and multiple crystals. Hot-pressed compounds, first introduced by Eastman Kodak as Irtrans, have been used, even though they generate some scatter. They can be made as large as the hot presses can hold. Recent advances have generated large crystals made by chemical vapor deposition (CVD).

10.2 Selection Criteria

The relative importance of the properties of materials to be considered depends on the application. One set of criteria applies to windows that do not form images but provide a protective shield for the system, and another to the elements inside, usually lenses. There are some considerations to be applied if the lenses are to be cooled to cryogenic temperatures or heated to scorching temperatures, or must undergo relatively wide operational temperature swings. Thus, for windows, one considers the following properties, in the following order: transparency as a must, strength and hardness for protection, thermal expansion, thermal conductivity, specific heat for high-temperature applications, refractive index for reflection loss, and sometimes rain-erosion resistance.

For a lens, one considers first of all its transparency; the refractive index and its change with wavelength and temperature (sometimes called the thermorefractive coefficient) are next. The thermal properties are secondary, usually because the inside of the system is under control. Hardness is to be considered for manufacture and less importantly for scratch resistance.

10.3 Reflection and Transmission

The standard equation for specular reflection from a single, relatively smooth surface at normal incidence is due to Fresnel:

$$\rho = \frac{(n-1)^2}{(n+1)^2} . \tag{10.1}$$

Of course, the refractive index value varies with wavelength. The more complicated Fresnel expression for nonnormal incidence is plotted in Fig. 10-1.

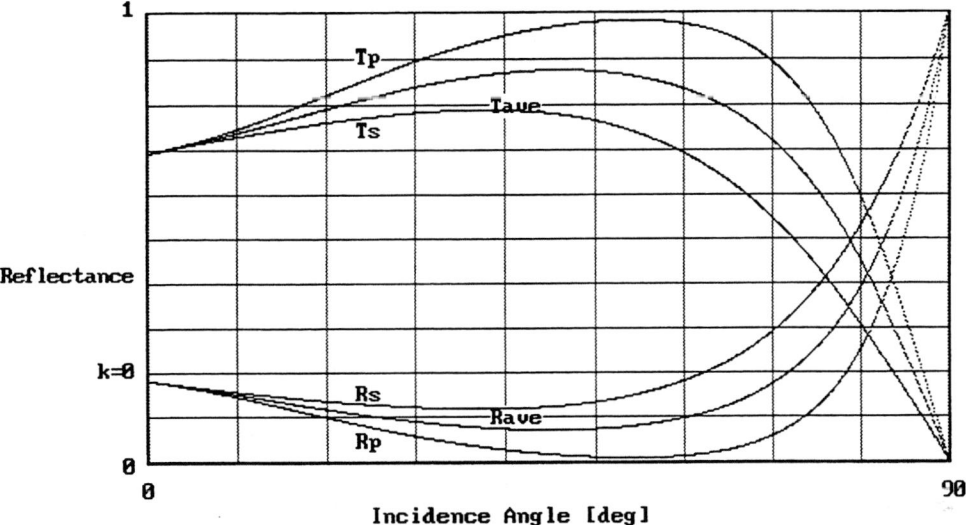

Figure 10-1. Fresnel reflectance and transmittance

The transmission and reflection from a plane parallel plate due to multiple internal reflections are

$$\tau_\infty = \frac{(1-\rho)^2 \tau}{1-\rho^2 \tau^2} \tag{10.2}$$

and

$$\rho_\infty = \rho + \frac{(1-\rho)^2 \rho \tau^2}{1-\rho^2 \tau^2} , \tag{10.3}$$

where τ is the internal transmittance, the ratio of the flux at the inside of the second surface to that at the inside of the first surface. This unmeasurable quantity is given by Beer's law of exponential attenuation. These equations are plotted in Fig. 10-2.

Note that when there is no absorption, $\tau = 1$ and

$$\tau_\infty = \frac{2n}{n^2+1} \ .$$

(10.4)

This function is plotted in Fig. 10-3. It emphasizes the increased reflectivity with refractive index, a problem that often occurs with infrared materials (like Si and Ge).

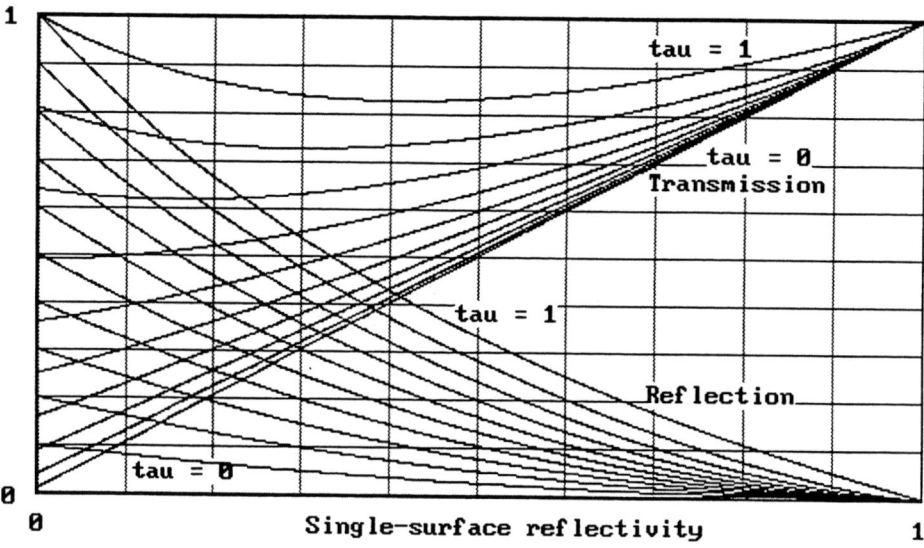

Figure 10-2. Transmission and reflection

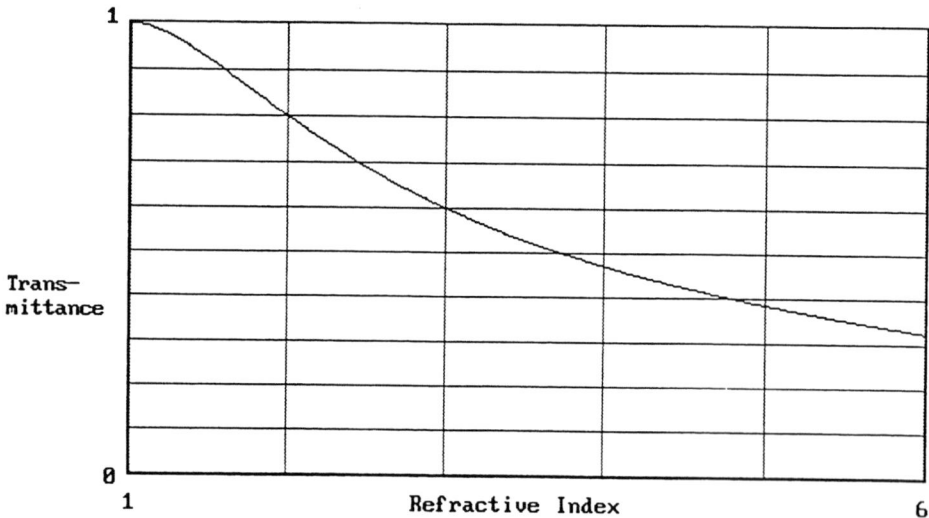

Figure 10-3. Transmittance vs. refractive index

10.4 Absorption

The two main absorption mechanisms are lattice vibrations and the generation of free electrons. When the frequency of the incoming radiation matches that of the lattice dipole, the lattice takes up the energy in the radiation as a mechanical vibration, thereby increasing the temperature of the material. The frequency of the dipole is given by

$$\sigma = \sqrt{\frac{k}{m}}, \tag{10.5}$$

where k is the binding constant and m is the reduced mass. One such dipole might consist of the positive charge of sodium and the negative charge of chlorine in salt. The wavelength of the resonance is given by the reciprocal of this. It is then clear that if the resonance wavelength is to be large, say 25 µm, the reduced mass must be large and the binding constant small. This small binding constant gives rise to the weakness, softness, chemical attack, and other undesirable properties of these long-wave materials.

   The thermal agitation of the lattice can increase the probability that electrons are freed from the valence bands to the conduction band. The probability that this happens is given by

$$P = e^{\frac{U-U_g}{kT}}, \tag{10.6}$$

where U is the thermal energy, U_g is the bandgap energy, k is the Boltzmann constant, and T is the absolute temperature.

10.5 Refractive Materials

Table 10-1 lists the most-used materials and some of their salient properties. The first column is the material, listed by either its chemical name or its familiar moniker. The next column is the spectral region of usefulness (in µm). This region must be taken with a grain of NaCl: it can only be given for a certain thickness and certain transmittance. The next column is the refractive index (dimensionless). This is not precise enough for lens design, but is an indication of the requirement for antireflection coatings. Density is given in terms of specific gravity, which is dimensionless. Then come thermal conductivity, thermal expansion, and specific heat capacity in the units shown. Hardness is given in grams per deciliter, a sort of percent, since water has a density of 0.01 gram per deciliter. Size is in inches, and comments are given in four-letter words.

   The first group consists of refractory materials that are useful as domes in the MWIR. They are hard and strong, have high melting temperatures, come in sufficient sizes, and are even relatively resistant to rain erosion.

   The second group consists of fluorides. They are useful in a variety of applications. However, LiF suffers from severe thermal shock. When polished too fast, it will crack. A material that has good thermal shock resistance has high thermal conductivity to conduct the input heat to its frame, high thermal capacity so that it takes

Table 10-1. Properties of Selected Infrared Materials

Material	$\Delta\lambda$ [μm]	n	ρ	T [C]	Hard	Sol %	Size in.	Comments
SiO_2	.2-3.4	1.44	2.2	1710	461	no	9	dome
TiO_2	.2–4.2	2.38	4.25	1825	900	no	9	dome
$Al_2 O_3$	.2–4.5	1.8	3.98	2030	1370	no	9	dome
MgO	.3–7.5	1.7	3.58	2800	692	no	9	dome
Spinel	.3–4.5	1.7	3.61	2030	1140	no	9	dome
Alon	.2–7	1.7		2200	1912	no	9	dome
LiF	.12–5.5	1.36	2.64	870	102	.27	9	shocking
CaF_2	.15–6.5	1.42	3.18	1360	158	no	9	cleavage
BaF_2	.2–10	1.44	4.83	1280	82	.17	6	dual lasers
MgF_2	.4–12	1.38	3.18	1255	576	.008	6.5	Irtran
NaCl	.2–14	1.52	2.16	801	14	36	12	salt
KBr	.3–25	1.49	2.75	730	6	54	12	
KRS–5	.5–35	2.34	7.37	414	40	.05	12	inhomo
CsBr	.2–33	1.64	4.44	636	20	124	7.5	slurp
CsI	.3–42	1.71	4.53	621		44	5.5	
KCl	.2–15		1.99	776	8	35	12	lo abs
Si	1.2–15	3.40	2.33	1420	1150	no	12	hi temp
Ge	1.8–15	4.00	5.33	936		no	12	lo temp
ZnSe	.6–24	2.23	5.27	400	178	no	24	Irtran
ZnS	.6–24	2.42	4.09	300	137	no	24	Cleartran
CdS	.6–20	2.33	4.82	1500	122	no		new
Se	.5–20	2.78	4.8	35		no		rotates
Glass	.3–2.5	1.55				no	36	
As–S	.5–12	⁻2		200	109	no	12	
$CaAl_2O_3$	.3–5	⁻1.7		800		no	12	
Amtir	1–14	⁻2		200	150	no	12	TI 1120
C	.22–80	2.41	3.51	3500	Moh 10	no	.5	
TPX	.3–80	1.55				no		machinable
Poly	.3–80	1.55				no		sheets

a lot of heat to increase the temperature, low thermal expansion so that even with some temperature increase the material creates little stress and a high strength so that it can take the stress. CaF_2 exhibits unusual cleavage, uninteresting cleavage. It tends to split along cleavage planes. Care must be taken in preparation and handling. BaF is a nice material for transmitting both HeNe and CO_2 lasers, in spite of the fact that the spectral range is said to extend only to 10 μm. It also has a very low refractive index. KCl has remarkably low absorption and is therefore good for windows for high-energy lasers.

The next group consists of long-wave materials. Sodium chloride, common table salt, has a high solubility in water, and is therefore attacked by it. KI and KBr are even worse. All of these have relatively low hardness and strength values. These are all caused by the low binding constant that permits long wavelength transmission. KRS-5 is a mixed crystal, a mixture of TlBr and TlI. Although it has good transmission to fairly long wavelengths and low water solubility, it sometimes suffers from inhomogeneity.

Semiconductors are next. Probably the most-used materials for lenses are Ge and Si. They have good (but not great) transmission through the LWIR, but both have high refractive indices, thereby requiring antireflection coatings, and they suffer from loss of transmission at high temperatures. This is due to the generation of free electrons. ZnS and ZnSe are two new materials that are best made by the CVD technique. The hot-pressed versions have scatter. They really do cover a wide spectral region, and although new and expensive, they can solve problems previous materials could not. GaAs is a very new material and it seems to have good properties, but it is still too soon to see it in commercial quantities.

Glasses consist of silicate glass, here mostly for reference, aluminates, and chalcogenides. Watch out for the water band in the aluminates. Watch out for inhomogeneity in the chalcogenides and lack of full transmission in the LWIR—also problems with high-temperature operation.

The final group consists of the really long-wave materials, diamond and polyethylene. Diamond is hard and strong, with a reasonable index, no water solubility, and all that good stuff. It is a wonderful material with really only one exception. Although gentlemen may say it costs too much, ladies have the right answer: it doesn't come big enough. It is a homopolar material; there are no dipoles for absorption.

A very exciting development in the materials area is the development of diamondlike coatings. A thin coating of the diamond form of carbon is put on top of any of the other materials to make a hard, impervious, fully transparent window. Germanium has been so coated for some tactical applications. The chief problem is that the coatings scatter more than is desired.

10.6 Mirrors

Mirrors are characterized by their surface reflectivity and polish and by the properties of the blanks upon which these polishes are established. Figure 10-4 shows some of these. The most important materials for thin-layer coatings are gold, silver, and aluminum. Of these, silver tarnishes, but gold is a noble metal and aluminum takes on an oxide. One form of this aluminum oxide is sapphire, a hard, strong, transparent overcoat. Gold has a reflectivity of 98%, while aluminum is only 96%. This 2% difference in the reflection is not very important, but for space systems that suffer from their own emission, gold has half the emissivity of aluminum.

Table 10-2 shows the properties of most mirror blank materials. They can be classified as "glasses," metals, and composites. The glasses consist of Pyrex, quartz (fused silica), ULE, and Zerodur. This list is in approximate order of decreasing coefficient of thermal expansion (CTE). Very large Pyrex mirrors have been made (8-m diameter) in the 1990s by the use of a rotating furnace. They are made with an egg-crate structure that permits the flow of a temperature-stabilizing fluid to pass through it. Quartz mirrors have been the choice for many applications, although recently the so-

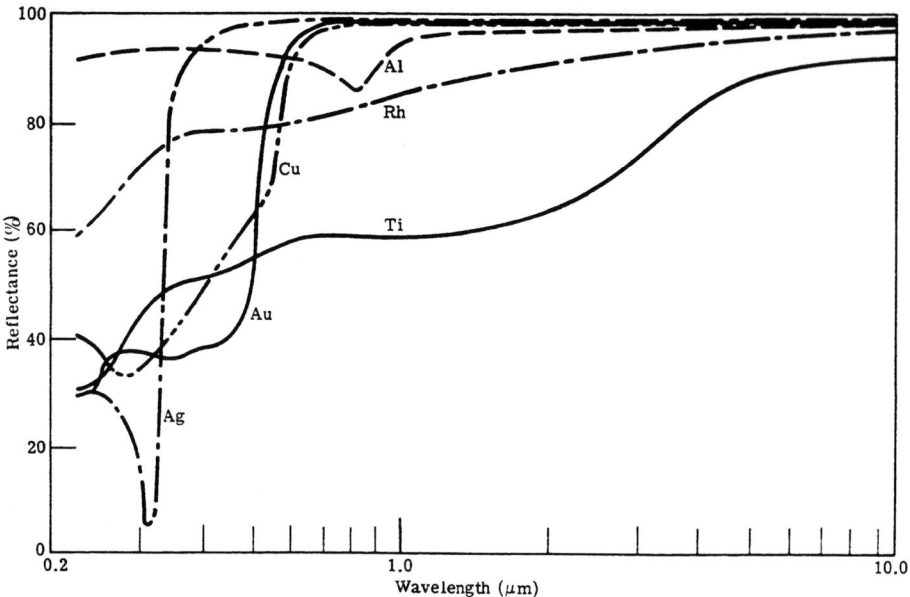

Figure 10-4. Mirror surfaces

called zero-expansion materials have largely displaced them. These materials have a zero CTE at only one temperature; it is incumbent upon the designer to determine the total expansion over his range of use. All of these "glass" mirrors take a very nice polish, but they must be mounted in special structures.

Metal mirrors consist of aluminum, beryllium, titanium, steel. and probably other metals, but only the first two need be considered. Aluminum is the choice for workhorse systems. Beryllium should be considered for applications that require extreme lightness. It has the highest strength-to-weight ratio of all except graphite epoxy. Beryllium is also the only established material that can withstand nuclear radiation—because it has such a low atomic number. However, no coating may be applied; it must be polished "bare." Unfortunately, beryllium, which is a hot-pressed compact, does not take a really good polish. For nonnuclear-radiation applications, the standard approach is a nickel coating applied chemically to a thickness of a few millimeters. Gold or aluminum is then applied as an overcoat. Aluminum can also be polished "bare," but a better result is usually obtained with an evaporated aluminum overcoat. Gold-on-nickel is also possible. The temperature range of use and storage should also be considered in matching, as well as possible, the CTE of the coating to the substrate.

The composites, graphite epoxy (GrEp) and SXA, have attractive properties for mirror blanks. GrEp has the highest strength-to-weight ratio of anything in the table. SXA, an aluminum matrix, has a CTE that can be tailored to match its environment and any other structure to which it mates—over a reasonable range. Both must be considered as potential outgassers in the space environment.

SiC has potential as a good, lightweight material that can survive the nuclear environment, but it is still an untried material, with no substantial pedigree.

Table 10-2 Mirror Blank Properties

Material	ρ	E	E/ρ	CTE	k	Z	Comment
	–	GPa	1/GPa	ppm/K	W/cm/K	–	!$@*&
Glasses							
Pyrex	2.23	65.5	29	3.3	0.011		CTE
Quartz	2.19	72	33	0.50	0.014		OK
ULE	2.21	67	30	0.02	0.013		not zero
Zerodur	2.53	92	36	-0.09	0.016		not zero
Metals							
Al	2.70	68	25	23.6	1.67	27	sprong
Be	1.85	287	155	11.4	1.93	9	polish
Composites							
SiC	3.15	403	128	3.4	0.88	40	new
GrEp	1.21	255	210	4.0	?		gassy
SXA	2.90	117	40	12.4	1.24		new

10.7 Glass vs. Metals and Composites

Glass mirrors take a better finish than the others, require special mounts, and are usually heavier, although they can be tested for residual strain (by birefringence). Metal mirrors can be made with integral mounts ("ears"), are lighter, can be mounted on an optical bench of the same material (for automatic temperature compensation), but do not take as good a polish and cannot be tested for residual strain. Anneal and anneal and anneal and pray. Composites are much like metals, but they have the additional problem of outgassing. Heat in a vacuum and heat in a vacuum and heat in a vacuum and pray. Note that anything that is outgassed will try to deposit on the coldest available surface. In space, this is either the detector array or the mirrors!

11 SOME OPTICAL SYSTEMS

Lenses and mirrors both are used in the infrared, and they have certain advantages and disadvantages. Mirrors are achromatic, testable, available in large sizes, can be readily athermalized, but they almost always get in their own way, because of reflection. In this section several of the more useful forms are reviewed to provide the reader with some reasonable starting points in the designs. The order progresses from single mirrors (in the sense of mirrors with power) through three-mirror systems, to catadioptrics, those with correctors. Then an exploration of variations on the themes.[1]

11.1 Single-Mirror Systems

These include the Newtonian, Herschellian, Pfundian and an eccentric pupil version of the Herschellian. The paraboloid is the right geometric shape for obtaining on-axis imagery that is free from third order aberrations. So it is used for infinite objects.

Herschel simply tilted the paraboloid, as shown in Fig. 11-1, to get the detector out of the way of the incoming beam. Of course, since it is tilted off axis, it suffers from coma and astigmatism. However, it is an extremely simple optical system that has no obscuration and can be set up in the laboratory or in the field .

A different version is to use only half the mirror (or less), as shown in Fig. 11-2 . This makes it an eccentric pupil. It stays on axis, but uses only about half the available aperture, so in a sense it has half the speed it could have—or twice the curvature. The normal way to make such a mirror

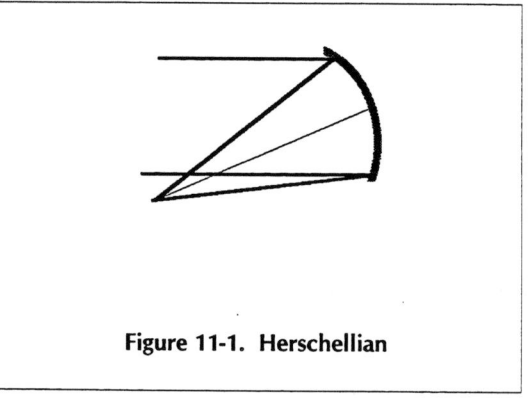

Figure 11-1. Herschellian

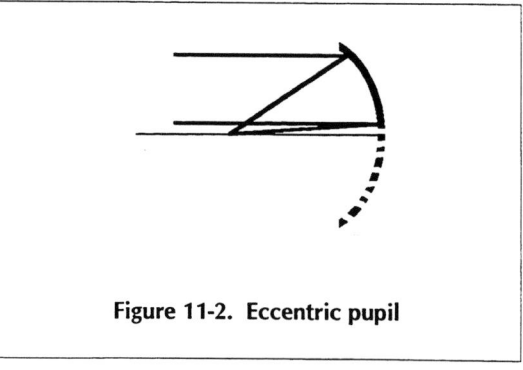

Figure 11-2. Eccentric pupil

[1] K. Schwarzschild, *Investigations in Geometrical Optics in the Theory of Reflector Telescopes,* Astronomical Mitteilung der König. Stern. zu Gottingen, 1905.

today is either to make the parent and "cookie cut" about three segments out of it, or to use numerically controlled diamond turners or polishers to form the off-axis figure. "Cookie cutting" is basically the use of a hole saw, using an appropriate grinding tool.

The Newtonian system, show in Fig. 11-3, also uses a paraboloid as the primary, but incorporates a folding flat mirror to get access to the detector. It is on axis, but it is highly obscured. The amount of obscuration is a function of the speed of the system. One can arrange the system so that the flat touches the top of the primary and is at 45 degrees. This is "closest packing." Then the system will be $F/1$, ignoring the curvature of the mirror and requiring the focus to be just at the edge of the beam, but it will be completely obscured. The vertical distance must always be just a little bigger than $D/2$; the center of the folding flat must be $f-D/2$ from the mirror. Therefore the linear obscuration will be $(2f-D)/4f$. The real obscuration will be the square of this and is actually a little more complicated because the flat is not symmetric.

The Pfundian in a sense is an inverse Newtonian; it applies the same principle of a folding mirror, but in this case the mirror comes first, and the obscuration is quite small, as shown in Fig. 11-4. The price that is paid is a larger flat mirror, but this is a relatively small price.

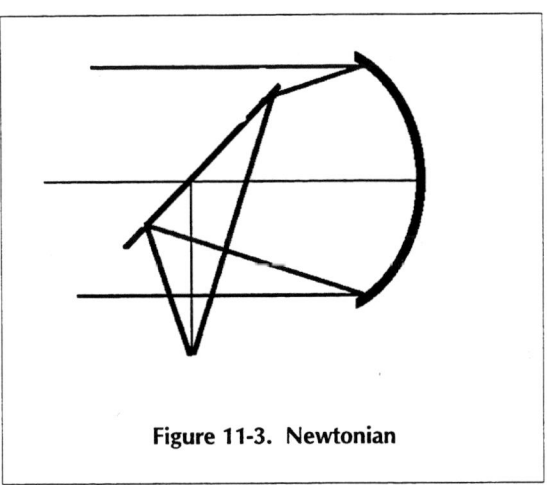

Figure 11-3. Newtonian

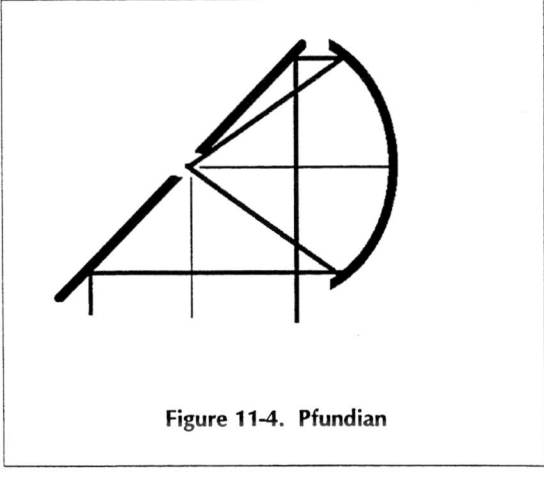

Figure 11-4. Pfundian

Each of these systems is inherently a *paraboloidal* mirror and suffers the aberrations of a simple paraboloid—on axis and off axis.

11.2 Two-Mirror Systems[2,3]

There are more opportunities to invent design forms with two mirrors that have power, and there are more forms. The systems include the Cassegrain family, Gregorians, and Schwarzschild's system.

A true Cassegrain[4] consists of a paraboloidal primary (so that the primary image on axis will have no third-order aberrations) and a hyperboloidal secondary. It is shown in Fig. 11-5. The hyperboloid has the property that all rays aimed at its primary focus, the focus of one sheet of the hyperboloid, will be reflected to the secondary focal point and will do it without aberration on axis. Another version of the Cassegrain, "Cass" to those who know it well, is with a spherical secondary. This is a

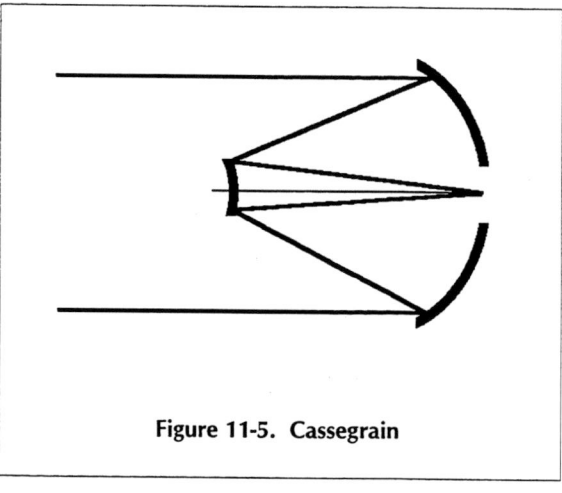

Figure 11-5. Cassegrain

form called the Dahl-Kirkham; it has somewhat less performance, but is considerably easier to make. The primary or both mirrors can also be made spherical. These forms provide even less performance, but are even cheaper to make. The Ritchey-Chrétien consists of two higher-order aspheres, provides better performance, but is more expensive.

The treatment by Gascoigne[2] that is cited above shows how these different forms fit together. It separated into two terms. The first can be eliminated by the use of a parabola; one then works to minimize the rest. However, one can minimize both terms by the use of special aspheres, and this is the route of the Ritchey-Chrétien. Of course, it is a more expensive telescope to make, as testified by the Hubble, which is an "R-C". So one can use a "Cass," go cheap with a Dahl-Kirkham, or go first class, and use a Ritchey-Chrétien. Ya pays yer money and ya takes yer cherce!

---

[2] S. C. B. Gascoigne, "Recent advances in astronomical optics," *Applied Optics,* **12**, 1419 (1973).

[3] W. B. Wetherell and M. P. Rimmer, "General analysis of aplanatic Cassegrain, Gregorian and Schwarzschild telescopes," *Applied Optics,* **11**, 2817 (1972).

[4] C. L. Wyman and D. Korsch, "Aplanatic two-mirror telescope: a systematic study. 1: Cassegrain configuration," *Applied Optics,* **13**, 2064 (1974).

The Gregorian[5] uses an ellipsoid rather than a hyperboloid as the secondary element, and it uses the well-known property of conjugate foci of the ellipsoid that light emanating from one focus will go to the other focus. It is shown in Fig. 11-6. Compared to an equivalent Cassegrain, it is longer, but it has a primary focus and has two concave surfaces, which are easier to test and therefore make. Gregorians are not used much because of their relative length, but the principle of having available a primary focus is used in modern systems, as eccentric pupil forms, as we shall soon see (with the MICOM system).

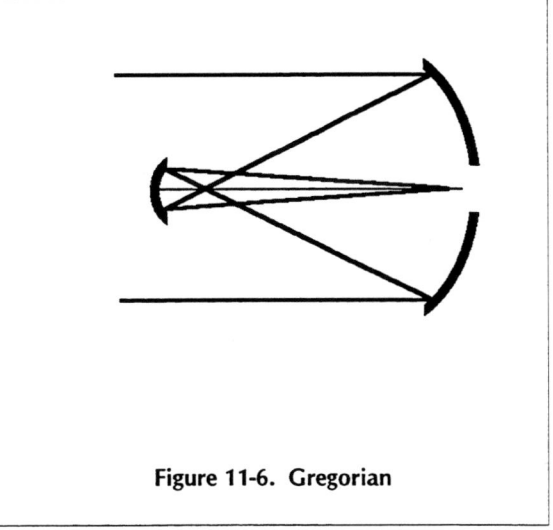

Figure 11-6. Gregorian

The Schwarzschild[6] is the first of the systems described so far that has a good wide field; however, apheres ruin the symmetry for relatively large field angles, the essence of good, wide-angle performance. Now consider the eccentric form of the system, as shown in Fig. 11-8. Most of the obscuration is gone. It has formed the bases of design variations for wide-field, compact systems.

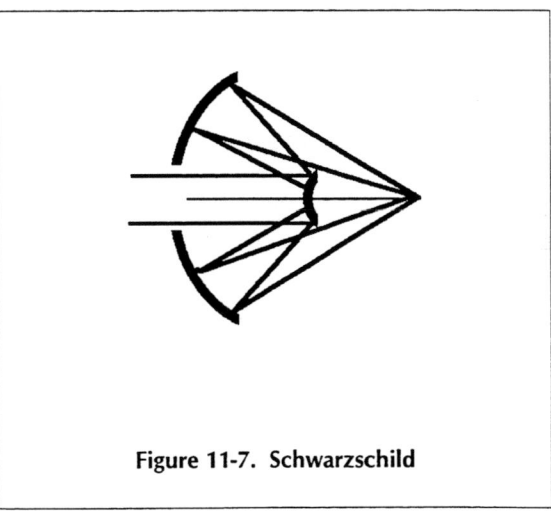

Figure 11-7. Schwarzschild

[5] C. L. Wyman and D. Korsch, "Aplanatic two-mirror telescope: a systematic study. 2: the Gregorian," *Applied Optics,* **13**, 2402 (1974).

[6] C. L. Wyman and D. Korsch, "Aplanatic two-mirror telescope: a systematic study. 3: the Schwarzschild-Couder configuration," *Applied Optics,* **14**, 992 (1975).

11.3 Three-Mirror Systems

Three-mirror systems fold the light back in the "wrong" direction. In fact, any odd number of mirrors will do this. Sometimes the wrong direction is right, and always an additional fold with a flat mirror will solve the problem.

The Korsch system, shown in Fig. 11-9, might be called a transverse three-mirror system for fairly obvious reasons. Most of these systems are readily realizable in slow speeds. Figure 11-9 also shows how the fold mirror obscures the beam for the transverse pass. It is a copy of the figure in Korsch's article; I think it is clear that a higher speed version will have considerably more obscuration. The obscuration is on the same order as for a Newtonian. Korsch has also written a nice article involving closed-form solutions for three-mirror systems.[7]

The Meinel-Shack system, Fig. 11-10, provides decent resolution over a reasonable field, is an on-line system, but also has considerable obscuration. The illustration shows a fairly

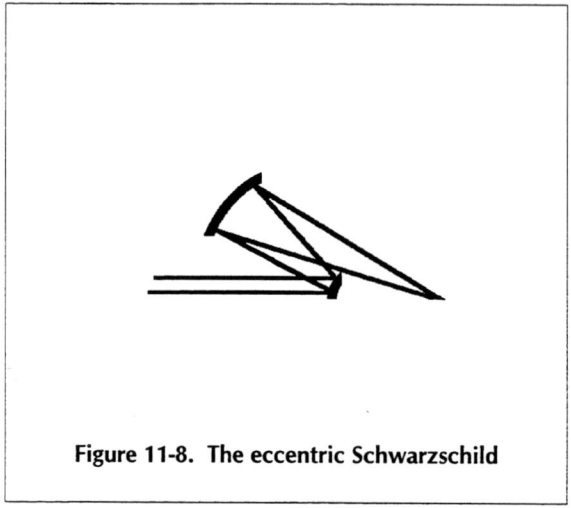

Figure 11-8. The eccentric Schwarzschild

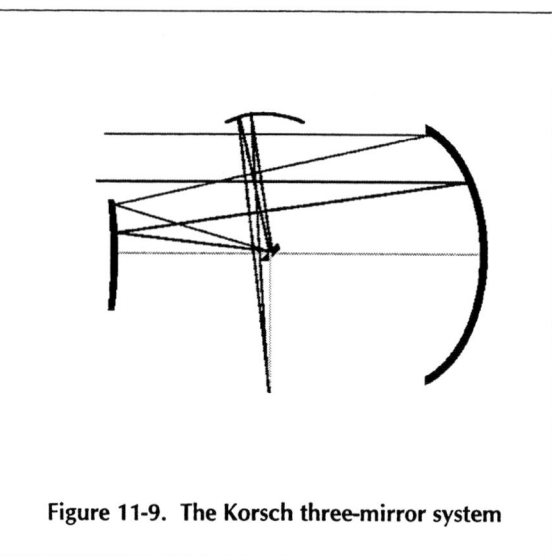

Figure 11-9. The Korsch three-mirror system

high-speed form. The F number is between 2 and 3. It is surely a more effective system when the speed is lower, and there is more space for the components. Recall that the slower system makes the aberrations lower but the radiometry of extended objects

---

[7] D. Korsch, "Closed form solution for three-mirror telescopes, corrected for spherical aberration, coma, astigmatism, and field curvature," *Applied Optics,* **12,** 2986 (1972).

worse. There is a message here: multiple-mirror systems tend to get in their own way and obscure themselves. The solution is to use eccentric pupils.

The system shown in Fig. 11-11 is an eccentric-pupil, reimaged configuration that was designed and built for the U. S. Army Missile Command (MICOM) in Huntsville, Alabama. The primary mirror is a 2-m-diameter sphere. It is tilted so that its axis passes through the prime focus F_1. In this sense it is like a Gregorian.

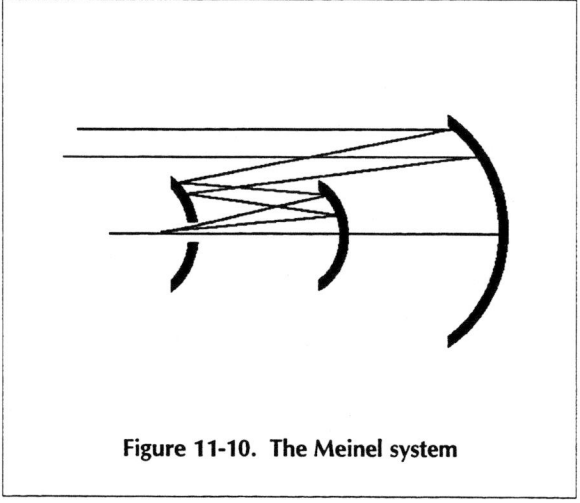

Figure 11-10. The Meinel system

The secondary mirror is an off-axis higher-order asphere that almost collimates the light toward the tertiary, which is a very similar mirror. It, in turn, brings the light to the final focus. The sphere was made from Zerodur in approximately one month. Each of the other mirrors took about six months as a result of their more complicated figures. They were made as parents, and then cookie-cut. The system was designed and used for measuring the laser radar cross section of various objects.

Stray light was an important issue in this design. Thus, a small flat was placed by F_1 so that the return light could be folded out of the plane of the figure and duplicate secondaries and tertiaries could be used. This meant that the outgoing and incoming beams were *slightly* offset in angle. But more importantly, their scatter was not important. Only the scatter of the primary was an issue.

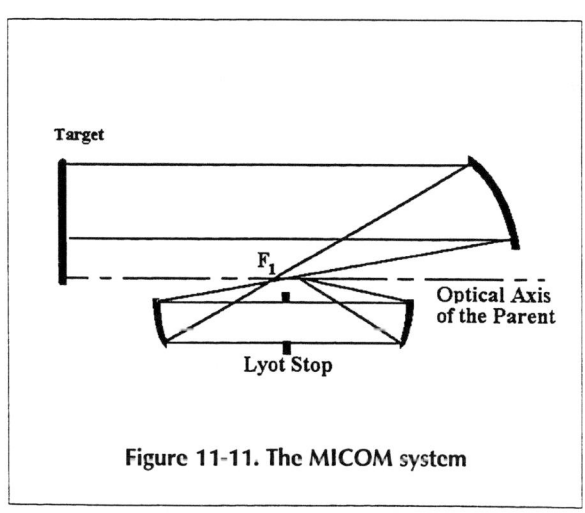

Figure 11-11. The MICOM system

The targets did not fill all the beam. Thus, a considerable amount of light could pass by the target and reflect from the back wall of the lab, thereby ruining the entire measurement. Accordingly, a stop was placed where a Lyot stop is indicated on the figure and imaged at the target. This shaped the beam. Such a stop was probably first used by B. Lyot to measure solar corona without eclipses. He used a circular mask that just covered the sun; the corona was then visible just outside this occulting disk. It is an important weapon in the optical arsenal of devices that require good scatter-rejection properties. The usual design procedure is to place the Lyot stop conjugate to the front of the baffle tube in front of the optical system.

11.3.1 The Walrus[8] and Seal[9]

I believe Walrus stands for wide-angle, large, reflective, unobscured system. As can be seen, it is in a Z configuration, makes use of the eccentric pupil concept, and depending on your definitions is a high-performance wide-angle system. In Fig. 11-12 the light comes from several field points at the bottom of the figure and is finally focused at these three field positions in the field stop (on the detector). The primary is convex; the secondary mirror is an ellipsoid; and the tertiary is a sphere.

The Seal, or spherical elliptical, all-reflective long, is a smaller Walrus! Shown in the same orientation, the light enters to a convex spherical primary and then to a prolate ellipsoid mirror that operates both as a secondary and tertiary, and the one mirror is on axis. The author reports slightly better performance than the Walrus with about half the length.

There is surely no reason that the ellipsoid could not be replaced by two eccentric mirrors to obtain a little more design freedom and probably better performance, but the double use of the one on-axis mirror is clever indeed. One can also scale down the system to all spheres, or use a conic convex mirror. These are the things of which new designs are made.

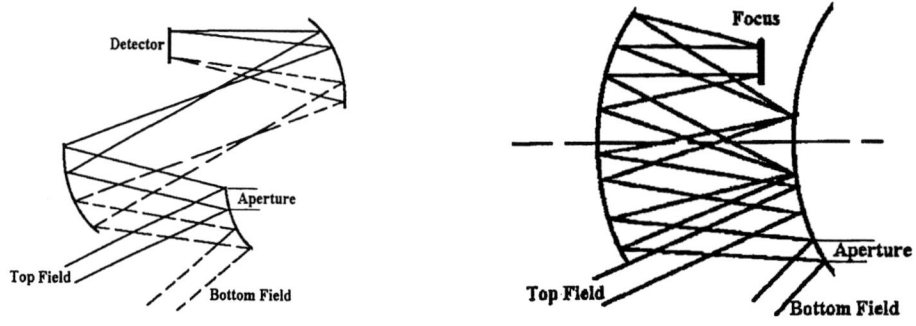

Figure 11-12. The Walrus and the Seal

---

[8] K. L. Hallum, B. J. Howell and M. E. Wilson, Wide Angle Flat Field Telescope, U.S. Patent 4598981 (1986).

[9] R. Calvin Owen, Baird Corporation, "Easily fabricated wide angle telescope," Proc. SPIE **1354**, 430 (1990).

11.3.2 Three-Mirror Anastigmats

Several versions of these systems exist. One by Cook,[10] Fig. 11-13, is an unobscured system; another by Cook[10] is *F*/3, and uses three conic mirrors (also) to cover a strip field of 10 degrees (see Fig. 11-14). Tolerances are typically 0.005 degrees for tilt and 0.00005 in. for decenter and spacing. Korsch[11] has published a similar design.

Other variations are possible.[12] One version is an *F*/2.6, 400 μrad, 30-degree system. To go further, one must set his own parameters and restrictions to get a real design. Performance will be better if the speed is decreased. Better resolution can be obtained over smaller fields, but the resolution can be optimized for the field size. It does not scale directly.

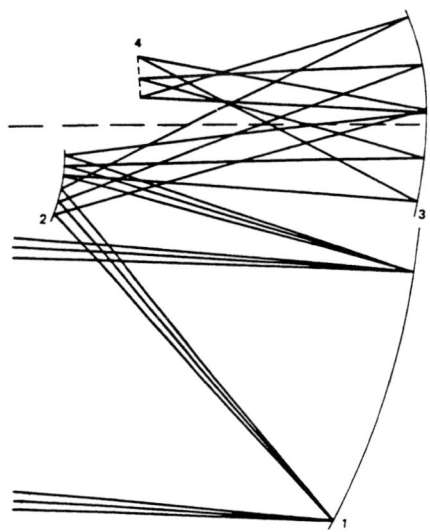

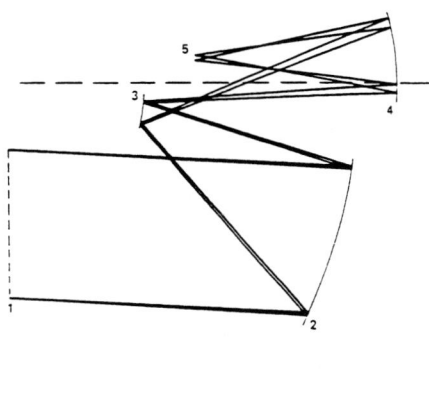

Figure 11-13.   **Figure 11-14.**

[10] L. G. Cook, "Wide field of view three-mirror anastigmat (TMA) employing spherical secondary and tertiary mirrors," Proc. SPIE **183**, 207 (1979) and US Patent 4,265,510 (1981).

[11] D. Korsch, Wide-Field Three-Mirror Collimator, Patent 4,737,021 (1988).

[12] L. Jones, Reflective and Catadioptric Objectives, Chap. 18 in *Handbook of Optics*, Volume II, M. Bass, E. Van Stryland, D. Williams, and W. Wolfe, Editors, McGraw-Hill, 1995.

11.4 Schmidt Systems

B. Schmidt in 1932 published
a note entitled *Ein lichtstarkes
komafreies Speigelsystem*, a
telescope system that now
bears his name. It consists of a
spherical mirror with a stop at
the center of curvature, the so-
called Schmidt principle. It is
shown in Fig. 11-15. All rays
are on axis; there can be no
coma or astigmatism. The only
third-order aberrations left are
field curvature, distortion, and
spherical aberration. The
spherical aberration is corrected by the use of an

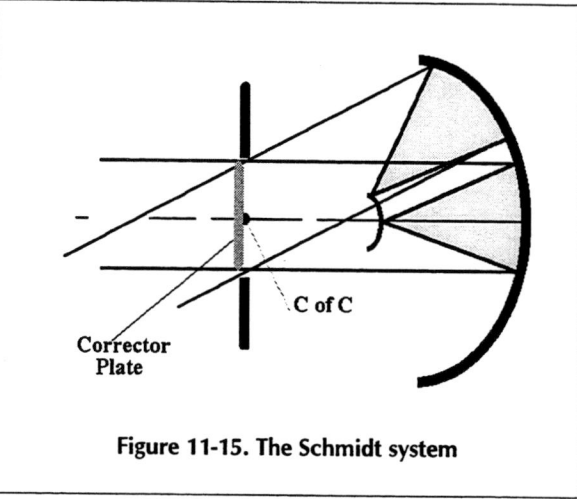

Figure 11-15. The Schmidt system

aspheric corrector placed at the center of curvature. A simple way to think about this is
that it predistorts the wave so that the spherical mirror looks like a parabola. It may also
be thought of as a phase delay or optical path difference that is a function of beam
height.

Bowen treats Schmidts, mostly for visible and astronomical purposes, and cites
other references that are more detailed. Linfoot's book[13] is probably the best of these;
the others are good for historical purposes. Bowen gives equations for the corrector
plate, for aberrations off axis, and for achromatizing and more.

Systems that have flat fields, using a refracting second element, have been
described by Baker[14] and Linfoot[15] They are sometimes called Baker-Schmidts or
Schmidt-Cassegrains.

All-reflective Schmidts have been studied by Abel.[16] There are several
configurations, involving the placement of a mirror corrector and sometimes an auxiliary
folding mirror. He also includes Baker-Schmidts.

---

[13] E. H. Linfoot, *Recent Advances in Optics,* Oxford University Press, 1955.

[14] J. G. Baker, *Journal of the American Philosophical Society,* **82**, 339 (1940).

[15] E. H. Linfoot, *Monthly Notices of the Royal Astronomical Society,* **104**, (1944).

[16] I. R. Abel and M. R. Hatch, "Wide field correction in reflective Schmidt systems by
a non-rotationally symmetric element," Proceedings SPIE **536**, 232 (1985) and **237**, 271
(1980).

11.5 Bouwers-Maksutov Systems[17]

During World War II the Russians, under the leadership of Maksutov, and the Dutch, under the leadership of Bouwers, both developed high-performance, wide-angle systems based on the Schmidt principle. They put the stop at the center of curvature of a spherical mirror. However, a corrector plate is **not** used at the aperture stop. Rather, a concentric, spherical correc- tor is used that is placed either in front of or behind the stop. They showed that this would correct for the spherical aberration of the primary by having their own spherical aberration of opposite sign. The system is then symmetrical for all angles. These systems may be called *front* and *rear* *Bouwers systems*. Both are shown in a single system in Fig. 11-16. Of course, the front system works better than the rear system, because it takes up more space. There

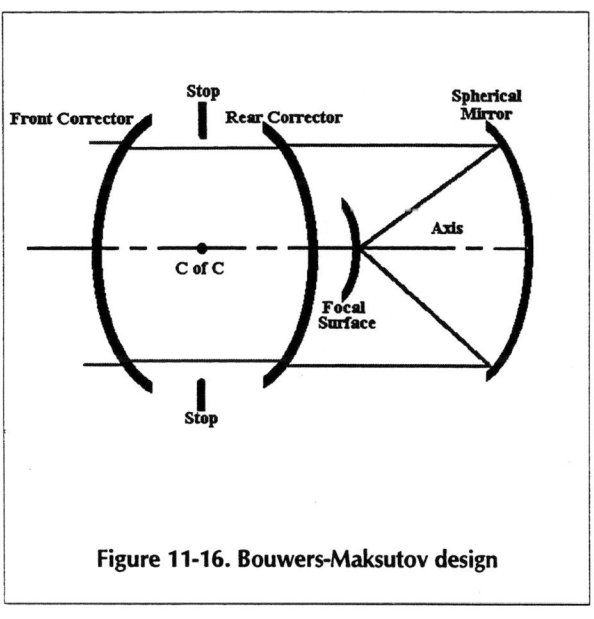

Figure 11-16. Bouwers-Maksutov design

is no such thing as a free lunch, a perfect vacuum, or an uncluttered horizontal surface! The Bouwers-Maksutov designs tend to get cumbersome in that the correctors get thick and have large curvatures—i.e. they are soup bowls! An additional variation is the Schmidt-Bouwers. In it, there can be either a front or rear corrector, or both, and a Schmidt plate at the center of curvature.

11.6 Reflecting Relays

Optical relays are used for several different purposes. They come in focal and afocal versions. They can relay stops for the purpose of controlling scatter. They can be used for controlling the speed angle of a detector array to limit the amount of unwanted background radiation, and they can be used to control the size of the beam. In an application described below, the Schmidt corrector is relayed. The design of lens relays has been brought to a fine art in telescope applications. Not much attention has been paid to reflecting relays. Here we discuss very briefly the use of afocal, mirror relays.

---

[17] A. C. S. van Heel, *Achievements in Optics*, North Holland, 1969.

One example of an afocal mirror relay is a clamshell relay system, shown in Fig. 11-17. For systems that do not have too large a field, one can place the foreoptics focus at the first mirror. The light is then spread to the second mirror, which reverses and collimates it. The first mirror then focuses the light back to the pole of the second mirror. The process can be repeated with another pair and another pair and another pair and another pair . . .

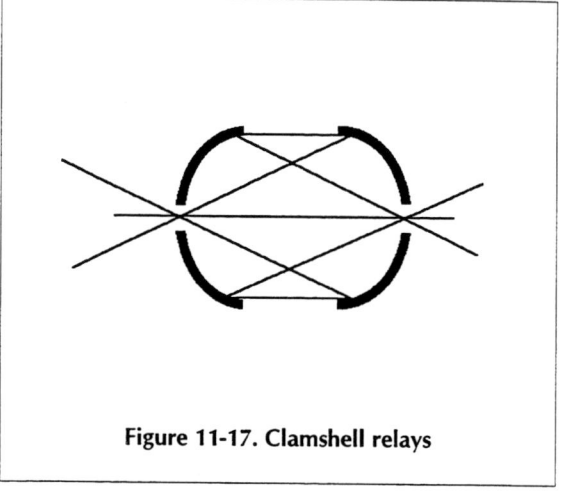

Figure 11-17. Clamshell relays

Afocal relays come in many guises, and they are often used in common-module FLIR applications in order to bring the beam of correct clear aperture onto a common module scanner of a given size. In every afocal relay, however, the Lagrange or Helmholtz invariant applies. For every reduction in aperture there is always a concomitant and proportional increase in the required field angle. This can be seen by reference to Fig. 11-18. Light is focused by a primary optic. It is to be relayed, but to do this, both the image and the F/cone must be captured by the

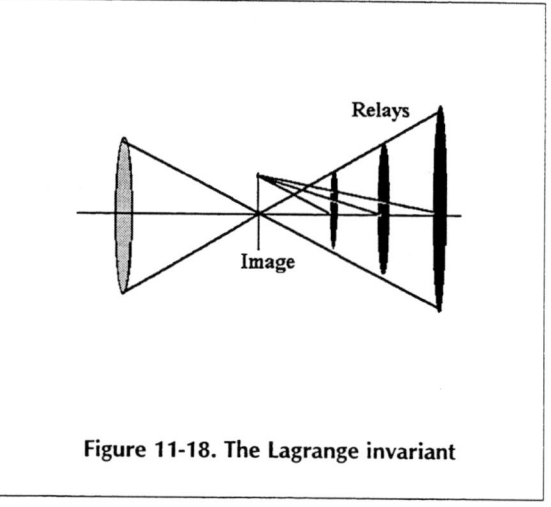

Figure 11-18. The Lagrange invariant

relay optics. It is readily seen that the closer candidates have small apertures, but the image has a large angular field for them. The more distant ones have larger apertures, but the image subtends a smaller angle. Similar triangles reign; all is proportional.

Wetherell has published a nice review on such devices.[18] They include the double parabolas, Mersennes, the Offner Cassegrain-parabola, and the double Cassegrain. These can be done as full systems or as eccentric pupils.

---

[18] W. Wetherell, "All-reflecting afocal telescopes," Proceedings SPIE **751**, 1989.

12 SCANNING SYSTEMS

The representative raster field of view was presented in Chap. 9. Sometimes this raster does not cover the entire area that is to be covered. The strip mapper was also discussed; this mapper can be generated by either a whiskbroom or a pushbroom scanner. The instantaneous field of view is the field that is covered by the optical system at any instant. It can have a single detector, a linear array, or a two-dimensional array as the defining field stop. The resolution element, reselm, or picture element, pixel, is the field subtended by a single detector element. The field of regard is the entire field that is covered by the system; it can be larger than the instantaneous field of view. It is also possible for the field of regard to be the same as the instantaneous field, or all three field can be identical. In trackers, the three fields are often identical. In staring systems with two-dimensional arrays, the instantaneous field and the regard field are identical, but there are many reselms in it. When a single detector is used to cover a raster, the reselm is the same as the instantaneous field, but the field of regard is the full frame. The instantaneous field and field of regard are usually specified in degrees, while the reselms are given as milliradians or microradians. These are illustrated in Fig. 12-1.

In this section various ways to scan the reselms or the instantaneous field over the field of regard are discussed. There is, as always, a tradeoff. Staring systems require very good, wide-field optics, and it has already been shown how much more difficult this is than very good, narrow-field optics. However, the scanners all require moving parts that are less reliable and require more space and more power. In some cases the starers also require choppers.

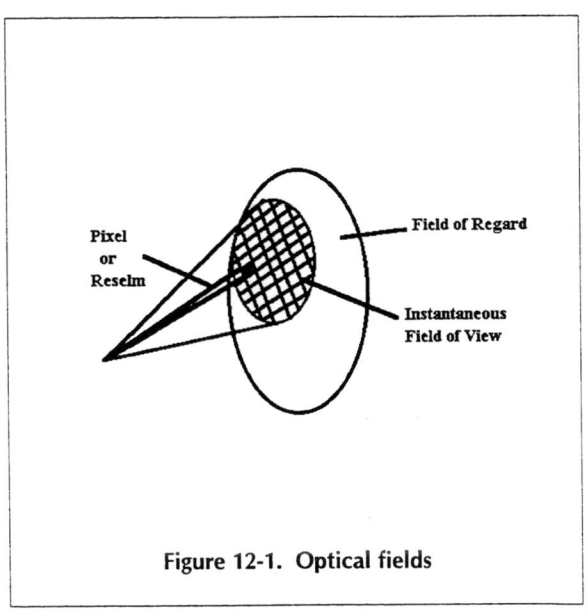

Figure 12-1. Optical fields

12.1 The Plane Parallel Plate

It is well known that a plate with plane and parallel faces displaces but does not deviate a ray of light. This can be used to generate a scan pattern and is shown in Fig. 12-2, with the plate in a converging beam. The plane parallel plate must be used in the converging beam, because if the plate is in the incident parallel beam it simply moves the beam up and down on the aperture, but does not change the position of focus. One *can* use the plane parallel plate in a converging beam as a scan mechanism. The equations can be derived from the refraction of a ray of light in a plane parallel plate. The ray is displaced, but it is not deviated. The equation for motion in the y direction is

$$y = \frac{\gamma t(n-1)}{n} \, ,$$

(12.1)

where γ is the tilt angle of the plate, t is its thickness, and n is its refractive index. One rocks the plate up and down in only the y direction (although if it were rocked in the x direction, the equations would be the same). The figure shows two parallel beams. Although one is shown larger than the other, this is only artist's license to differentiate them. The figure also shows the plate in two different positions and the resulting different positions of the focus of the system.

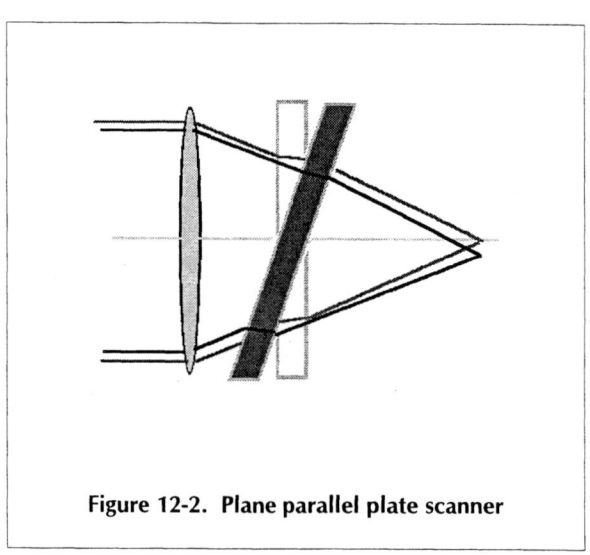

Figure 12-2. Plane parallel plate scanner

12.2 Refracting Polygon Scanners

A form of this scanner has been used successfully by the AGA, now AGEMA, infrared imagers. In their realization they use a refracting octagon to accomplish the (fast) line-scanning operation. In some realizations they also use the octagon for the vertical or frame-scanning operation, which is slower. In other instruments, they use an oscillating flat mirror for this (slower) scan.. The octagon may be considered as four plane parallel plates.

The system is shown in Fig. 12-3. Each pair of faces operates as a rotating plane parallel plate. They bring a portion of the field of regard, the frame, onto the detector or array, one pair at a time.

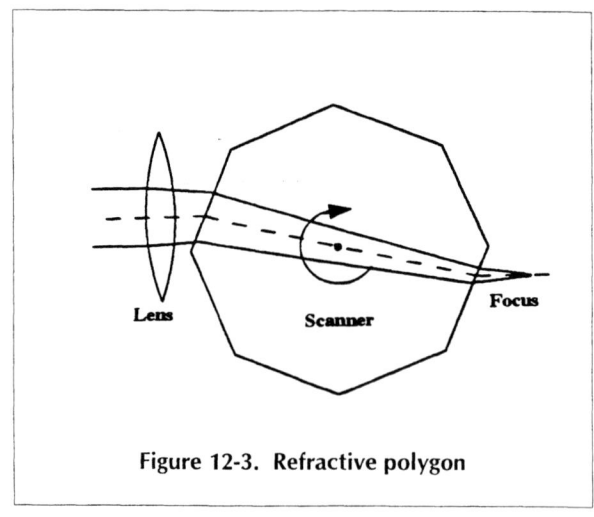

Figure 12-3. Refractive polygon

Care must be taken that the beam is blanked when the apexes of the octagon are in the field of the optical system, because light can then be refracted from two different directions, thereby generating what I call double-fielding or walleyeing. This is the restriction that limits the scan efficiency of the system. The usual practice is to electronically blank the system when such double-fielding does occur.

12.3 Risley Prism Scanners

Risley prisms were first used as a means of generating variable attenuation by translation of the prisms, one with respect to the other.[1] They can also be used to deviate a beam, singly and in different combinations. These rotating prisms are put in front of the objective of the optical system, as shown in Fig. 12-4, and they are capable of generating a host of different patterns, depending on the choice of prism angle, relative

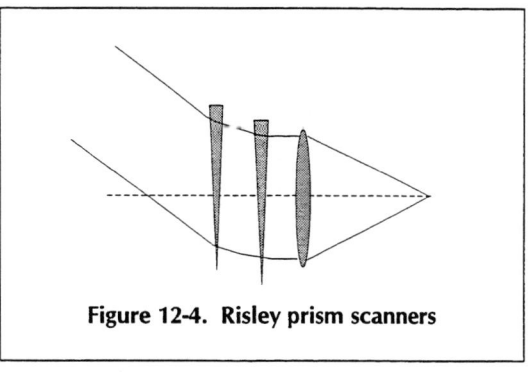

Figure 12-4. Risley prism scanners

rotation rates, phase difference in the rates, and direction of the rates. The deviation of a single prism δ is given by the simple formula for prism deviation:

$$\delta = \alpha(n-1) , \qquad (12.2)$$

where α is the prism angle and n is its refractive index. For two of them, it is just the sum, assuming that they are aligned. But these prisms are, in general constantly changing their relationships with each other. Thus, the angles must be defined in terms of some two-dimensional coordinate system. The deviations can be written

$$\delta_x = \delta_1\cos(\omega_1 t) + \delta_2\cos(\omega_2 t - \phi) . \qquad (12.3)$$

$$\delta_y = \delta_1\sin(\omega_1 t) + \delta_2\sin(\omega_2 t - \phi) . \qquad (12.4)$$

The individual prism deviations are denoted by δ_i, the rotation rate (which can be positive or negative) by ω_i, and the angular phase shift between the prism orientations by ϕ. Table 12-1 shows some of the patterns. The letter m represents the ratio of rotation rates (with sign); k is the ratio of prism deviation angles (prism wedge angles). The

---

[1] F. A. Jenkins and H. E. White, *Fundamental of Optics,* Second Edition, McGraw-Hill, 1950.

patterns are shown in Table 12-1 for various combinations. The ratio of the rotational frequencies of the two prisms is m; the ratio of the prism angles is k; the phase relation at the time they start is ϕ. For raster scans, the prisms are rotated at the same rate in the opposite direction. For rosettes, they are rotated oppositely at different rates.

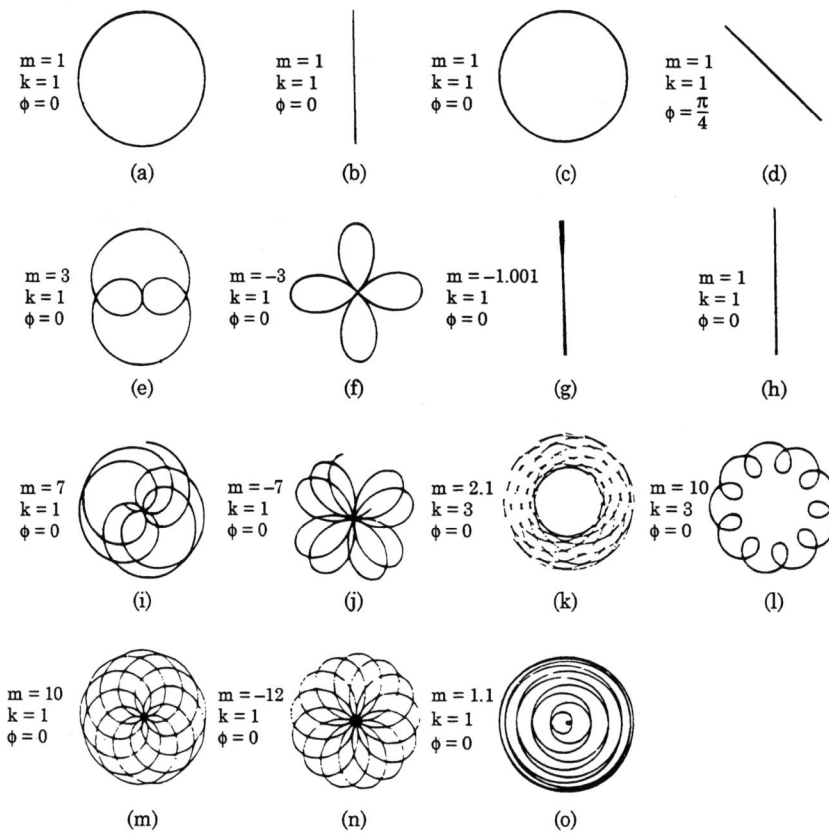

Table 12-1. Rotating prism scan patterns

12.4 Rotating Flats and Axe-Blade Reflecting Scanners

Some of the earliest scanners were designed for airplanes. They were flat mirrors tilted at about 45 degrees to the vertical and spun about their axes. Such scanners had, at best, a scan efficiency of 50% when viewing the ground. An improvement is to use a double-ended system, with two optical systems, as shown in Fig. 12-5. This could be used, for instance, in an interleaved way or for two different spectral regions. Note that both these scan methods generate a bow-tie pattern, and both have instantaneous fields that elongate with the scan angle at a rate different than they "widenate," and they rotate around their own center. This is more of a problem with arrays than with single detectors.

Another way to address the problem of limited scan efficiency is to use an axe blade—in a sense, to divide the mirror in half. Figure 12-6 shows this. Now the problem is double-fielding. Light should come from both sides of the field of view at the same time and get to the optical system and the detector. What a scramble that would be. So, as shown in Fig. 12-6, the objective must be made much smaller than the scanner. That isn't efficient either. One can use a four-sided axe blade, or pyramid, but the pyramid has the same problems of the axe blade, even though it covers only about 90 degrees. Image rotation also exists.

An analysis of the double-fielding may be made by viewing the system end on. Then the positioning of the aperture on the scanner must be such that the line, which represents the edge of the axe blade, must not cut the aperture during the active scan.[2]

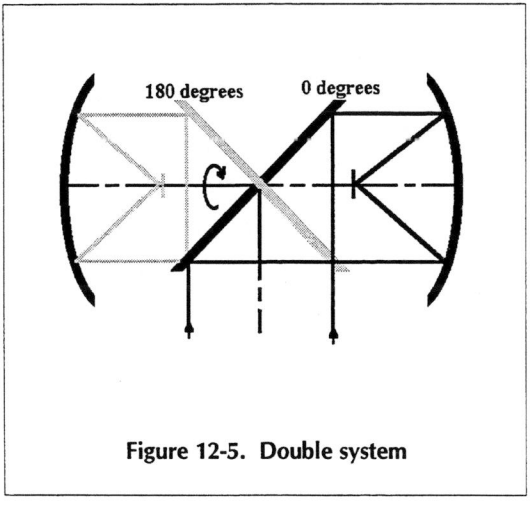

Figure 12-5. Double system

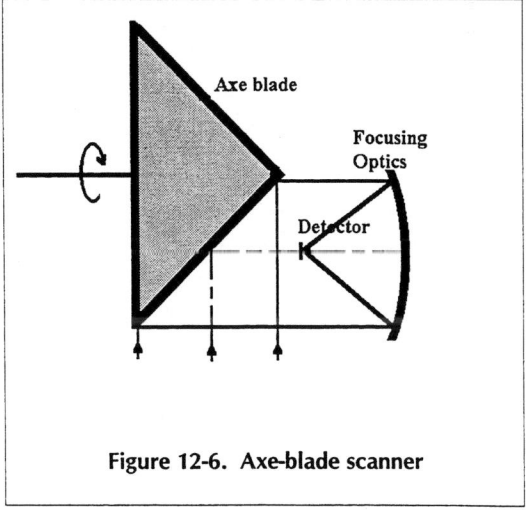

Figure 12-6. Axe-blade scanner

---

[2] W.L. Wolfe and G. J. Zissis, *The Infrared Handbook*, U.S. Government Printing Office, pp. 10–15; Available from SPIE, Bellingham, WA.

12.5 The Reflecting Polygon Scanner (Carousel)

The choice of a scanner for FLIRs (an acronym for forward- looking infrared, but often used to indicate any kind of real-time imager) has devolved largely to the polygon scanner, although Risleys still exist (and have some interesting advantages). A rotating polygon is shown in Fig. 12-7. The reflecting polygon may be considered the reflecting analog of the refracting polygon. It might be considered a bunch of mirrors, but not rotating about axes on their faces. Although reflecting, polygon (carousel) scanners are effective devices, they have their disadvantages that need to be considered. These might be called design challenges. There is beam walk. There is humping (a perfectly innocuous term that is described in the next section), because the projected area changes with scan angle. This can be reduced or eliminated with a properly sized and placed objective element. There is dead time that must exist during the time a transition is made from one facet to the next. Beam walk can be reduced by limiting the aperture size. The image does not rotate. The scan efficiency is not easy to analyze and neither is the position that the aperture must take for relatively wide scanning, but it can be done. Beiser and Johnson's chapter in *Handbook of Optics* is a good reference.[3]

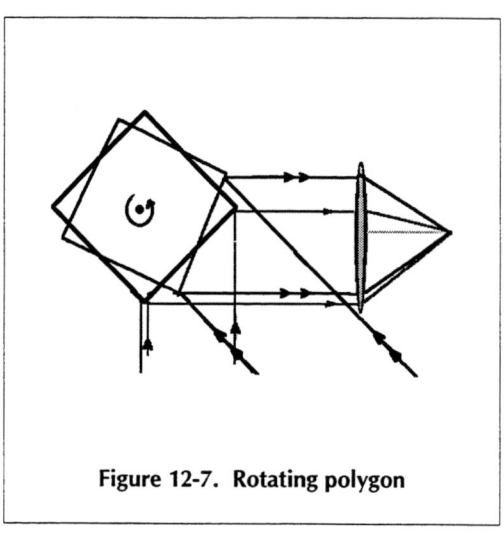

Figure 12-7. Rotating polygon

12.6 The Kennedy Split-Aperture Scanner

The choice for down-looking, strip-mapping applications seems to have been made; it is the Kennedy scanner. As shown in Fig. 12-8, it consists of a rotating cube, although it can be a polygon of any number of sides, that divides the light to two different "outrigger" mirrors, which in turn direct the light to the objective (at the top). The quadragon is good for scanning 180 degrees, while a triangular prism is fine for 120 degrees. The second pair of outriggers is not necessary, but sometimes it is nice. As the cube rotates, the light going in one direction experiences a larger aperture by the $\cos\theta$ obliquity factor. Because the mirror face next to it is turned 90 degrees, its projected aperture goes as the sine of the scan angle. Thus, the effective aperture turns out to be $(\cos\theta + \sin\theta)A$, where θ is the scan angle and A is the area of a face. This generates a

[3] L. Beiser and R. Johnson, Scanners, Chapter 19 in *Handbook of Optics,* Volume II, edited by M. Bass, E, Van Stryland, D. Williams and W. Wolfe, Optical Society of America, Washington, DC, 1995.

"humping," a variation in the response of the system with respect to the angle of scan, the field angle. Too bad it didn't come out $\sin^2\theta + \cos^2\theta$! A second effect is beam walking. The beam on the outriggers moves up and down as a function of scan angle, and any inhomogeneities will result in scan noise.

12.7 Image-Space Scanners

The scanners described above scanned in object space. The scanning elements were in front of the entrance pupil of the optical system. In a sense, the optical system did not know anything at all about the fact that a different portion of the field of view was being viewed. This is not strictly true for the refractive scanner placed behind the objective, although that same scanner could have been in front and been a lot larger. The classical image-space scanner is the Nipkow scanner, that was used for early television, in the days before the television tube was available. This is shown in side view (except the Nipkow plate is tilted to show the holes) in Fig. 12-9. An appropriate optical system forms an image of the field of view on a disk that has a spiral of holes in it. This disk spins. A lens right behind it images the entire field of view on the single detector. If there were

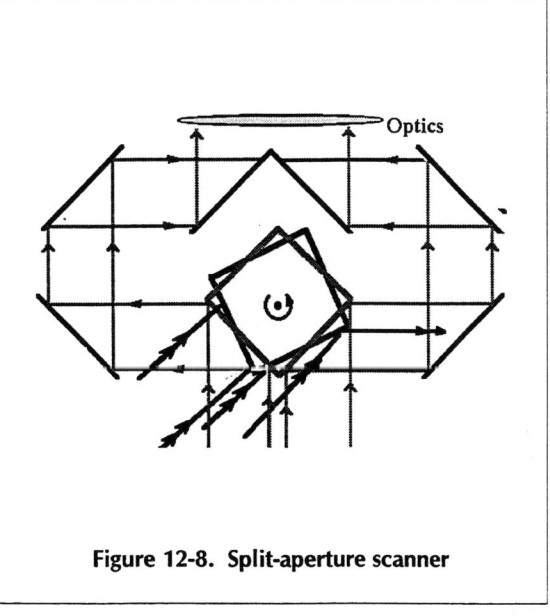

Figure 12-8. Split-aperture scanner

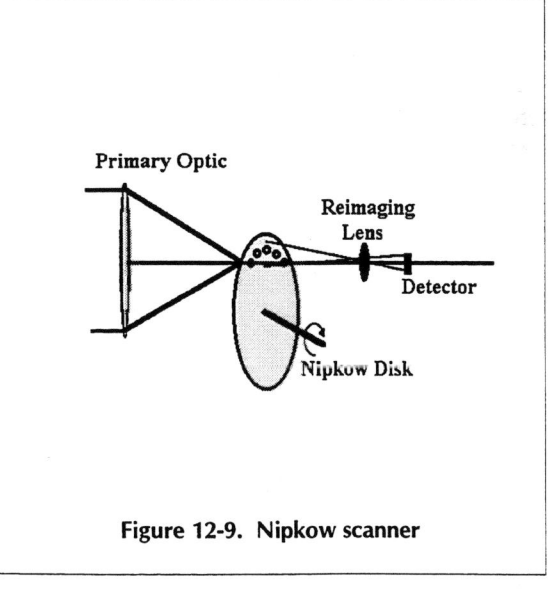

Figure 12-9. Nipkow scanner

no disk, all of the field would be imaged on the detector and sensed as a single spot. But the disk rotates, letting light through the one hole at a time that is in the field of view. A great deal of the onus of this system is on the relay lens that must have (de)magnification that puts the full field onto a small detector.

12.8 Narcissus

One of the problems with image scanning devices is a phenomenon called narcissus, after the vain Greek demigod. The effect arises from the scanning mirror moving an image of the detector (or detector array) across the detector itself. On axis, the detector sees itself, and off axis, it sees an image of the surround, which is warmer. The result is a band right down the middle of an image of this type, which is black in a white-hot system (and vice versa). The solutions are straightforward: Don't use image space scanning; don't use refractive optics; don't have refractive optics concave toward the detector; use highly antireflective coatings on the surfaces.

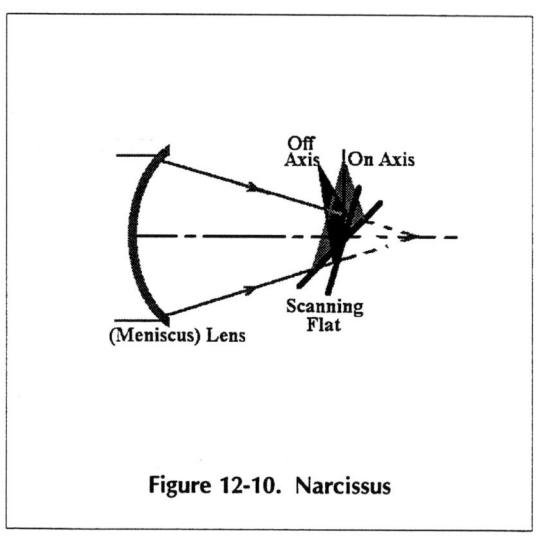

Figure 12-10. Narcissus

Narcissus can also occur in testing. A simulator for an infrared missile seeker should account for the relatively cool background of the sky. Often, as a result, the target is superimposed on a cold blackbody background, and both are required to be in a vacuum. The vacuum must have a window. The test system looks at the window, which has a finite reflectivity. Thus the system looks back at itself like our friend of the Greek myth. The solution is simple, if it has been considered: tilt the window a little . . . enough.

13 REAL-TIME IMAGER DESCRIPTIONS

Real-time imagers need to be described and characterized in a manner that has not been covered in the sensitivity analysis nor in the specifications of optical systems. Therefore, this chapter covers these descriptions and the extent of their applicability.

13.1 Contrast and Modulation

These two terms represent the difference in levels between the target and the background. In the visible, black to white has a contrast of 100%, and this seems to be an intuitive concept. Modulation is quite similar to contrast, but has a slightly different mathematical expression.

Contrast is defined as

$$\text{Contrast} = C = \frac{E_{max} - E_{min}}{E_{max}}, \tag{13.1}$$

where E represents the irradiance. It is clear that when the minimum is zero, the contrast is 100%, and when the maximum and minimum are equal, the contrast is zero. This is consistent with intuition.

Modulation is defined as

$$\text{Modulation} = M = \frac{E_{max} - E_{min}}{(E_{max} + E_{min})}. \tag{13.2}$$

The same two intuitive values apply. The two are related by

$$M = \frac{C}{C+1}. \tag{13.3}$$

13.2 Spatial Frequency

Temporal frequency is the reciprocal of time. If a cycle of a sine wave (or a square wave or a sawtooth or . . .) lasts for 0.1 s, then the frequency, the number of cycles in a second, is 10 cycles per second. This used to be abbreviated cps, but to honor Heinrich Hertz, it is now called hertz and abbreviated Hz.

Spatial frequency is the reciprocal of spatial distance. If a film can resolve 0.1 mm, then the spatial frequency should be 10 lines per mm. It has been common to specify the resolution of film in line *pairs* per millimeter, a spatial frequency. Note, however, that sneaky little *pairs* in the statement. It leads to a sometimes confusing factor of two. This occurs in other guises as well. In optics, a sine wave or a square wave has a black portion and a white portion in a single cycle. Thus, one line pair corresponds to a cycle. A line is a half cycle. The spatial frequency corresponding to the width of a line has a factor of two in it; the spatial frequency corresponding to a line pair does not.

Spatial frequencies can be specified on a linear basis, as indicated above. This is usually in lines or line pairs per millimeter, but can be in other linear units. The specification can also be in angular terms; this is usually in terms of cycles per milliradian. Note that cycles have no dimensions, so this can come out as reciprocal milliradians. A cycle can also be measured in radians—as there are 2π radians in a full cycle. Fortunately, this specification is almost never used.

13.3 Modulation Transfer Functions

The modulation transfer function (MTF) has been used to describe the resolution performance of optical systems for more than twenty-five years,[1] but to some it is still a little mysterious. Several different descriptions will be given in the hope that the reader will identify with at least one of them.

The performance of a circuit or an audio system is often described in terms of its transfer function. This is the ratio of the output voltage, current, or power to the input value as a function of frequency. Audio buffs have often required that their system have 3 dB down points at 20 Hz in the bass and 20,000 Hz in the treble. Although the electrical transfer function has both a modulus (amplitude) and phase, audio buffs do not worry too much about the phase function (which can lead to distortion). Recall that 3 dB means a factor of two. An optical system can be described in a similar fashion. However, the input and output values are the modulation, while the independent value is the spatial frequency. The optical transfer function is therefore the modulation ratio as a function of the spatial frequency. It has both an amplitude and a phase. The phase can distort the image—move it a little. The modulation transfer function is the modulus of the optical transfer function. It is nice to know that there is an analog to the relation between the transfer function and the impulse response. In electrical terms the transfer function is the Fourier transform of the impulse response. In optics, the MTF is the Fourier transform of the point source response, also called the point-spread function (PSF).

Another way to view this is in terms of the transfer of four-bar charts of different frequencies. Such a chart has four black bars and four white bars of equal width. Usually the bars are seven times as long as they are wide. Assume that an optical system has no trouble whatsoever in imaging crisply and cleanly bars that have a frequency of 1 line pair per millimeter. It would then have the same contrast (and modulation) at the output as at the input and a modulation transfer (MT) of 100%. The optical system may not do so well with a chart that has a frequency of 10 lp/mm (cycles/mm); its MT may only be 10%, because the optics smeared the lines as a result of aberrations or diffraction. At 200 cycles per mm, it may be down to 1%, and so on. A plot of all these transfer values

---

[1] J. W. Goodman, *Introduction to Fourier Optics*, McGraw-Hill, 1968; J. D. Gaskill, *Linear Systems, Fourier Transforms and Optics*, Wiley, 1978.

as a function of spatial frequency is the MTF. The MTF can be combined with the NETD to obtain other figures of merit.

13.4 Minimum Resolvable Temperature Difference[2]

The first step in defining a minimum resolvable temperature difference (MRTD) is to combine the concepts of NETD and MTF. The second step is to incorporate the fact that a real-time imager is viewed by an operator. First things first.

The NETD was defined as the minimum temperature that can be sensed from two targets that completely fill a pixel with an SNR of 1. This requires that the MTF is 100%, because the first target fills the subtense of the detector, and then the second target does the same. However, as the targets get smaller, the system works less well, and this can be accounted for by the MTF. The MRTD is the NETD divided by the MTF. As the targets get smaller, they must have larger temperature differences to be sensed. The MRTD will increase with increasing spatial frequency.

I have found that the reciprocal of the MRTD is a more useful figure of merit, because it follows the trend of the MTF. Unfortunately I may be a minority of one.

The second step is taking account of the properties of the viewer. The human eye has several properties that improve the performance over that predicted by the MRTD defined above. Perhaps the easiest to understand is that the eye has an integration time of about 0.1 second, while real-time viewers have frame times of 1/60 to 1/30 of a second. This means that the eye integrates from 3 to 6 frames, thereby improving performance by the square root of these factors. The eye also has properties of spatial integration, so that the lines in a test chart, which are typically seven times as long as they are wide, are more easily seen by a factor of the square root of seven. The eye also operates as a spatial matched filter, thereby reducing the noise. It is beyond the scope of this tutorial text to evaluate this quantitatively.

13.5 Limitations and Other Figures of Merit

The use of the MRTD, including the eye factors, explained why real-time imagers consistently worked better than predicted. The MRTD did not, however, relate to real images, since it corrected for the properties of bar charts. It also has the unfortunate characteristic that people are in the loop, and people are not consistent. It is suspected that some vendors hired engineers just because they had incredibly good visual perception. They were used for the MRTD tests!

Other charts were used, and some proposed real images. There is still no universally accepted standard, and those working in the area are not sure how to get there. I think it is because people are involved. People are no darned good, but some are better than others.

---

[2] J. M. Lloyd, *Thermal Imaging Systems,* Plenum Press, 1975.

13.6 MTFs for Diffraction-Limited Systems

It has been shown[1] that the MTF is the autocorrelation of the pupil function. A square aperture will have a linear MTF, as shown, and a circular aperture will have a more complicated one given by[3]

$$\text{MTF}(\xi) = \frac{2}{\pi}\left[\cos^{-1}\left(\frac{\xi}{\xi_0}\right) - \frac{\xi}{\xi_0}\sin\left(\cos^{-1}\left(\frac{\xi}{\xi_0}\right)\right)\right], \tag{13.4}$$

where the cutoff (spatial) frequency is given by

$$\xi_0 = \frac{1}{\lambda F} = \frac{2NA}{\lambda}. \tag{13.5}$$

The diffraction-limited MTFs for both the square and circular aperture are shown in Fig. 13-1. For estimation purposes, the linear MTF is very useful. It also can be seen that filling the detector with the Airy disk so that the diffraction-limited angular diameter is 2.44 λ/diameter gives about 50% modulation for a circular aperture and 60% for a rectangular aperture.

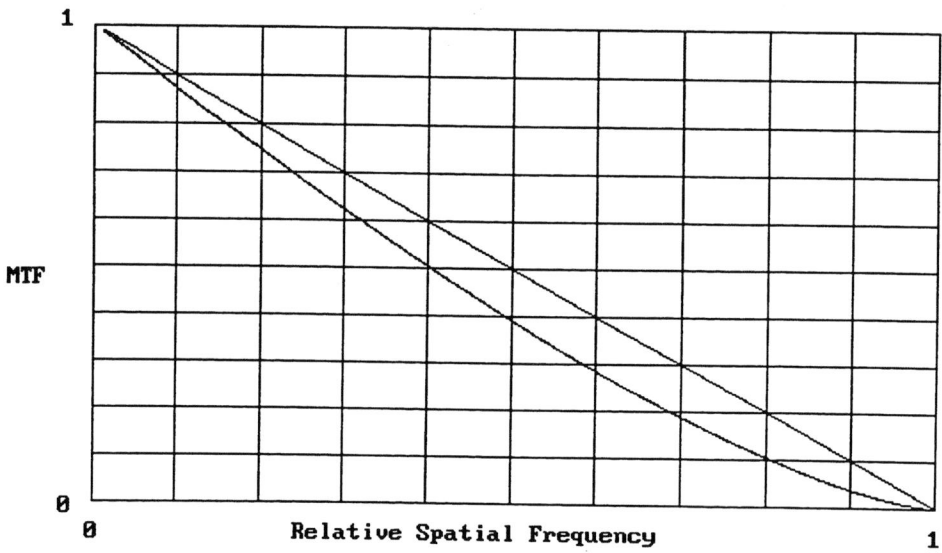

Figure 13-1. MTFs for diffraction-limited systems

---

[3] W. J. Smith, *Modern Optical Engineering,* McGraw-Hill, 1966.

14 DESIGN EXAMPLES

Several different design problems and approaches are presented in this chapter to bring together the concepts that have been presented thus far. The problems are necessarily brief and incomplete, but they show the approach and some of the tradeoffs.

14.1 The Design Process

Infrared system design is not a process of synthesis. Rather it is a process of invention, analysis, and iteration. The process I have found to be most productive is to analyze first the geometry—calculate the field of regard, the field of view, and the pixel size. Next calculate the time line—the frame time, dwell time, and the bandwidth. Then calculate the sensitivity on the idealized basis described above. The last step is the optical scheming. If all of this goes well (and it probably won't the first time) then repeat the sensitivity calculation with more accuracy and realism and do some real optical design.

We will apply these principles to the problems delineated below.

14.2 A Night-Driving Imager

The problem, stated simply, is to design an infrared device for driving a car in the dark and during times of poor visibility. This is a real problem. At least one company has hinted that it has such a project in place, and similar devices have been used by the Portland, Oregon, police.

What should the field of regard be? One answer is the full angular extent of the windshield as seen by the driver. This is approximately 90 degrees horizontally and 30 degrees vertically. Another answer, as we shall see, is that the field should be as much as you can get, perhaps with slewing. The field of view should be about 30 degrees, because that is about what we can get with a reasonable optical system. What should the angular resolution, the angular pixel size be? It should be the angle a man subtends at a reasonable stopping distance. We assume (and there are arguments based on the Johnson criteria[1] for this) that we should get five pixels across a man at a distance of 100 m. If the man is about average in size, each pixel will have to be about 5 cm. This requires a pixel subtense of 0.5 mrad. Then, since 30 degrees is approximately 0.5 radians, there are 1000×1000 pixels in the field of view. We assume that the system is a real-time imager, so that the frame time is 1/30 of a second. If a single detector were to be used, the dwell time would be 0.03 µs. We know that this is too short and will give a bandwidth in the many megahertz region. So we immediately plan for an array of some sort. The input SNR in the 8- to 12-µm region is about 10^7 (see Chap. 6). When this is divided by the square root of the bandwidth, we have idealized NETD values in the MWIR and LWIR regions, respectively, of 3.5 and 0.006 K for a staring system with a bandwidth of 15 Hz. One reason the MWIR looks so bad is that I assumed a quantum efficiency of 0.01 as in PtSi.

---

[1] L. M. Biberman, *Perception of Displayed Information,* Plenum Press, 1973.

Enter a little of the real world. The detector arrays of choice would be HgCdTe or PtSi. The former comes in arrays no larger than 512×512 and the latter has a quantum efficiency of about 0.01. One choice is to use the 512×512 array but increase the angular subtense. This is an iteration to make the specifications consonant with practicality. The use of PtSi provides a restriction to the MWIR and the low quantum efficiency (QE). The tradeoff, then, as can be seen in the computer printout of Appendix C, is between the uniform PtSi with relatively easy optics, and the very sensitive HgCdTe with much tougher optics. This is a tradeoff that will not be pursued further here. The equivalent design is done by unremarking the last line of the subroutine **Setup** in Appendix C. This changes the band and the QE.

Another option is the use of high-temperature HgCdTe. These detectors can be operated in the MWIR at about 200 K, a temperature that is obtainable with a Peltier cooler. Then the input D^* form of the equation must be used, as with thermal detectors, shown below. One could also consider an array that is 30 degrees high and scans the full width of 90 degrees. This would probably require a staggered array. For a single column, the bandwidth would be 500 pixels times 30 frames per second, divided by two, or 7500 Hz. TDI could be used, if necessary.

The systems just described must be cooled, and cooling is anathema. One might try the design with an array of thermal detectors that do not have to be cooled. Such arrays come only in numbers like 250×350, so that a compromise must be made. The sensitivity calculation is made according to the thermal detector menu.

$$\text{NETD}^{-1} = \frac{\pi}{4}\frac{\alpha D_o}{F}\frac{D^*}{B^{1/2}}\frac{\partial L}{\partial T}.$$

(14.1)

For calculation purposes, assume a resolution of 1 mrad, use a specific detectivity of 10^9, let the aperture diameter be 3 cm, the detector length and width be 0.005 cm, and find that the change in radiance with respect to temperature in the 8- to 12-μm region is 6.31×10^{-5}. Therefore, the NETD is about 170 mK. This could be a pretty good system, especially if the arrays can be made with more elements.

These three approaches have been briefly explored to show the design process. Much is left to be done to incorporate more realism, practicality, and engineering design. Other imagers can be designed along these same lines. The geometry for a medical scanner will be different, as will that for a device that is used for the security monitoring of a parking lot or for imaging the electrodes of a smelter. The principles, however, are the same.

14.3 ICBM Interceptors

In the heat of the cold war there was intense design work done on systems that could carry out the detection of ballistic missiles at launch, during midcourse, and in reentry. By way of illustration, one design is carried out through its basic steps.

Consider the interception of a ballistic missile in midcourse. The entire flight is about 30 minutes. Assume that one has about one minute of the flight to accomplish the task (perhaps because there are more to find). The uncertainty area provided to the sensor by some other means is a field of view of 10°×30°. One needs six detections in order to establish a trajectory; therefore the frame time is 10 s. To be on the safe side (and to make the problem a good example), assume that this is only 1 s. The required uncertainty in positional accuracy for this detection is 1 km. The range when the rocket first appears above the atmosphere can be as large as 5 Mm (1 Mm=1000 km). Therefore the required pixel size is 0.5 mrad (1 km divided by 5 Mm). Since 10° is 175 mrad, there are 350 pixels vertically and 1050 horizontally. We can assume a (vertical) linear array and parallel scanning, accomplished perhaps by a gimbal. Then, the bandwidth is 525 Hz. (I took the frame time of 1 s, found a dwell time of 1/1050 = 0.952 μs and used the Shannon theorem, as shown in Chap. 9). The geometry is done for now.

The SNR equation was given earlier as

$$\text{SNR} = \epsilon \tau_a \tau_o \eta^{1/2} g^{-1/2} \frac{A_s A_o}{R^2 \sqrt{A_d B}} \frac{L_q}{\sqrt{E_q}} , \qquad (14.2)$$

where all efficiency factors are weighted averages and the radiance and irradiance values are over the spectral band. We will assume that the pertinent emissivity-area product is 1 m². and that the smallest detector we can get is 25 μm and that the largest usable mirror diameter is 1 m. For now, ignore the efficiency factors—assume them to be unity (but we do use the emissivity-area product). Then for a 250-K target, the SNR is approximately (see the program in the Appendix entitled ICBM.BAS)

$$\text{SNR} = \frac{1.16x10^7}{\sqrt{E_q}} . \qquad (14.3)$$

The radiance is that from the target, while the irradiance is the photon flux density on the detector from all sources. This means that for an SNR of 10 the photon irradiance on the detector must be approximately 10^{12}. The issue is to see what it takes to do this. We must protect the system from the sun's radiation, from the Earth's radiation, and from its own emission. The solar irradiance is calculated from the radiance of a 5900 K blackbody and the solid angle the sun subtends at the Earth. The photon irradiance in the spectral band is 2.87×10^{16}. This is the irradiance on the aperture. Even if the sun is not in the field of view, this radiation will be scattered into the system. A baffle must be used to attenuate this by a factor of about 2×10^7. Such an attenuation is possible, but the design is beyond the scope of this text.

The Earth can be considered a blackbody at 300 K (the surface) and one at 250 K (the clouds) on a fifty-fifty basis. It subtends a projected solid angle of approximately π. It is a reasonable approximation, but in the detailed design, it needs to be done more thoroughly. In fact, I will just assume that the Earth is a blackbody at 275 K. The irradiance is 5.46×10^{17}. There is more irradiance on the aperture from the Earth than from the sun—because the assumed spectral band reaches to long wavelengths and the solid angle is so much larger.

The calculation of the self-emission is based on the radiance from each element and the amount that each following element passes—either by transmission for refractors or reflection for reflectors. I will assume that the system has a well-defined solid angle of acceptance defined by a focal ratio (F number) of 3 and that the optics consists of three mirrors with reflectivities of 0.95. Then the irradiance is given by

$$E_q = \left[(1-\rho_1)L_q(T_1)+\rho_1(1-\rho_2)L_q(T_2)+\rho_1\rho_2(1-\rho_3)L_q(T_3)\right]\Omega \ , \qquad (14.4)$$

where the counting proceeds from the detector on out toward the aperture. If the system is of uniform temperature and the mirrors all have the same properties, the equation simplifies to

$$E_q = (1-\rho)(1+\rho+\rho^2)\Omega L_q(T) \ . \qquad (14.5)$$

This is also evaluated in the ICBM.BAS program. For the required photon irradiance, the optical system must be kept at about 46 K.

Other evaluations include the effects of the moon, stars, the galactic background, and any other celestial sources that might be in or near the field of view. Stars are special cases because they are point sources and can be confused with the target. Techniques using spectral discrimination or trajectory analysis must be used.

The launch detector is designed in a similar fashion, but the geometry, background, and source are all different. The source is a point source, but it is the radiance of the plume of the rocket as it takes off. It has a tremendous amount of flux, much of which is concentrated in gaseous emission bands; much is also blackbody radiation. Most of this is attenuated by the atmosphere, so atmospheric transmission calculations are of great importance. The background is the top of the atmosphere and clouds, and this is not a uniform background. Discrimination will be of the essence. One would like to look from geosynchronous orbit to attain constant surveillance, and that means the resolution will be limited.

The reader may wish to explore the program for this example, which is given in the Appendix C.

14.4 A *Peacekeeper* Detector

Once upon a time, in the long, long ago, the USA was in a cold war with the USSR. President Ronald Reagan named our massive retaliatory ICBMs *Peacekeepers,* and it was proposed that they be housed in underground tunnels and moved around as if they were part of a shell game. We may thank the good Lord that the cold war is over and that this is no longer necessary. We should fervently pray that circumstances do not require a renewal of this effort! However, the detection by any opponent of our *Peacekeepers* makes a good illustration.

We postulate that the tunnels are in an area that is 200 km by 200 km, and that the tunnels are each 20 km long. The location of a missile can be ascertained because it causes a 1 K temperature increase in a spot that is 10 m^2 in area. We may assume that the earth is a blackbody at 300 K. This increase in temperature may be caused by the engines, motors, and power supplies that keep the missile at the ready. The area and temperature have been chosen to be reasonable and to make the problem challenging!

There seem to be three options for the vehicle that carries the infrared sensor: an airplane, say, at 30 km altitude, a geosynchronous satellite at 35 Mm altitude, and a low-altitude satellite at 100 km.

We can eliminate the two extremes almost immediately. The aircraft will be in considerable danger. The geosynchronous satellite is at an altitude of 35 km and will require a pixel angle of less than 0.3 μrad, which in turn at 12 μm will require an aperture diameter of almost 100 m!

We are left with the satellite at 100 km altitude. A satellite like that travels at about 7 km per second (90 min to travel the 24,000-mile circumference of the Earth). The full field of view is 90 degrees. The maximum slant range is 141 km, and the required angular pixel size is calculated at the edge is 176 μrad. There are, therefore, $45 \times \pi/(180 \times 1.76 \times 10^{-6}) = 4463$ pixels in a line. The bandwidth may be calculated from the formula, or step by step. The line time is 1.76×10^{-6} rad $\times 100$ km/ 7 km per sec = 2.51×10^{-5} s, and the dwell time is 5.62 ns. Oh my! This requires that there must be enough detectors in either a pushbroom or whiskbroom mode to increase the dwell time to the time constant of the fastest detectors. For the pushbroom, it is done automatically. For the whiskbroom, it has to be almost 180.

Looking ahead, we note that it will be hard to design a pushbroom with that large an angular field, and it will be hard to design a whiskbroom with that spin rate. But let us consider the SNR now, since infrared design is an iterative process.

We use the imaging form of the SNR equation, and 300 K. Now we know, however, that the reciprocal NETDs are 1.16×10^3 and 8.10×10^3 per root Hz in the MWIR and LWIR, respectively. Since the line time is also the dwell time for a pushbroom, the values are about 5.88 and 40.9, which give NETDs of 170 and 24 mK. The LWIR result gives us the required sensitivity with an SNR of about 4, but the MWIR result is unacceptable. Notice, however, that this required the use of 4463 detectors. The whiskbroom design will require the same number of detectors, but in a different configuration. This is because it really doesn't matter whether you increase the parallelism or use TDI to obtain the added sensitivity. We can even use the whiskbroom detectors in a square array of 66×66.

The pushbroom is an extremely demanding optical system. A field of view of 90 degrees with a resolution of 170 μrad cannot be done. The optical design is too difficult. Suppose the system is divided into three; that may do it. First, calculate the diffraction limit. At 12 μm the diameter for this resolution must be 17 cm. Okay. The spherical aberration will be 289 μrad (for an *F/*3 mirror). Asperize. The off-axis angle for one of the three is 15 degrees. The coma will be 1.83 mrad, and the astigmatism will be 11 mrad. It seems a long reach to correct the designs enough to satisfy the requirements, but maybe. One of the three-mirror anastigmat designs described above for a strip field or an all-reflective Schmidt might do it.

The same restrictions apply to spherical aberration and diffraction for the whiskbroom. Coma will be reduced by 4430/66 = 66 to an acceptable level. Astigmatism will be reduced by the square of this, or 4430, to about 2.5 μrad, a lovely result. Efficiency factors will drive the number of detectors up, for with a single, flat rotating scanning mirror, or even a Kennedy system, the scan efficiency is going to be about 0.25, the transmission about 0.4, the fill factor about 0.5, and the QE about 0.8. This results in an overall efficiency of 4.4% (using the proper square roots and $g = 4$). The Kennedy scanner does not rotate the array, although it adds size. The mirror should probably be four-sided, and the efficiency will go up. Atmospheric transmission has not been taken into account.

The next steps include calculating the size of the scanner, which is surely more than 17 cm, its scan efficiency, and other details. The principles have again been shown, and they can be applied to other requirements for strip mappers like the hot-rolled strip of steel, remote sensing scanners like the thematic mapper and its cousins, and even medical, whole-body scanners.

14.5 Summary

There is a similarity to the design approaches in all these cases. As was pointed out in the introduction, the process is one of iteration. First, one considers the geometry, then the bandwidth and time-constant requirements. Then one can use the simplified idealized equations for sensitivity calculations, using an appropriate spectral band. The choice of that band comes from experience and some thoughts about the general nature of the target radiation and the background. The first calculation of optical performance can be done with the diffraction calculations. This is the "ball park" start of the design process. However, even with the availability of sophisticated design programs, it makes a great deal of sense to take these simple steps. Many companies now have detailed programs that start with such details as the solid-state properties of the detectors and the MTFs of the optics; these are useful in later stages of the design process. As with most computer design approaches, one must keep the physics and reality in mind. Computers do not lie, but they can deceive.

Appendix A: FIGURE PROGRAMS

A number of figures in the text have been generated by QuickBasic. This Appendix provides the source code for them. There is some documentation via remarks in the earlier programs, but I felt it soon became redundant and stopped.

The first program listing calculates blackbody curves in four guises, although only two of them are in the text. The first (Fig. 4-6) is the spectral radiant emittance as a function of wavelength with the Wien distribution curve also shown. The second (Fig. 4-7) is the blackbody curve, but as a function of frequency, i.e., wave number. The next two are the logarithmic versions of these.

The second program is the radiation signal-to-noise ratio, the theoretical input for a photon-noise limited system (Fig. 8-1). It uses some of the same algorithms for obtaining the photon radiance and its temperature derivative.

The next is a plot of the Fresnel reflectance and transmittance curves for angles of incidence from 0 to 90 degrees, for a refractive index of 3.4 (silicon), and for no absorption. The values of real and imaginary parts of the index can be changed in the program once it is on your computer. This is Fig. 10-1.

Figure 10-2 shows the reflectance and transmittance values for a plate with multiple inter-reflections for a set of values of single-surface reflectances and internal transmittances.

Figure 10-3 is the same multiple-pass transmittance for a nonabsorbing sample for different values of refractive index from 1 to 6.

```
                                    'Blackbody Curves--Figures 4-6 and 4-7
GOSUB Setup
GOSUB Wavelengths
GOSUB Frequencies
GOSUB Logwaves
GOSUB Lognumber
Again: GOTO Again           'So that after the fig is done, the screen
                            'does not say, Press any key to Continue
END

Setup:
CLS
SCREEN 9
pi = 3.14159: c = 3E+10: h = 6.626E-34
c1 = 2 * pi * c ^ 2 * h * 1E+16: c2 = 14388
RETURN

Wavelengths:
SCREEN 9                              'set the screen for SVGA
VIEW (100, 40)-(600, 275)             'set an appropriate screen area
xmin = 0: xmax = 25                   'set extreme ordinate values
ymin = 0: ymax = .012                 'set extreme abscissa values
WINDOW (xmin, ymin)-(xmax, ymax)      'set appropriate calculational values
'insert captions and tick values
LOCATE 23, 15: PRINT "Figure 4-6.  Spectral radiant exitance vs wavelength"
LOCATE 22, 35: PRINT "Wavelength [um]"
LOCATE 8, 1: PRINT "Spectral"
LOCATE 9, 1: PRINT "Radiant"
LOCATE 10, 1: PRINT "Exitance"
LOCATE 11, 1: PRINT "[mW/cm2/um]"
LOCATE 20, 10: PRINT 0
LOCATE 13, 10: PRINT 5
LOCATE 6, 9: PRINT 10
LOCATE 21, 12: PRINT 0
LOCATE 21, 24: PRINT 5
LOCATE 21, 36: PRINT 10
LOCATE 21, 49: PRINT 15
LOCATE 21, 62: PRINT 20
LOCATE 21, 74: PRINT 25
LOCATE 18, 30: PRINT "300K"
LOCATE 8, 39: PRINT "380K"
'draw the grid lines
FOR x = xmin TO xmax STEP 1           'vertical lines
   LINE (x, ymin)-(x, ymax), 8        'the 8 gives gray grid lines
NEXT x
FOR y = ymin TO 1.1 * ymax STEP .001  'horizontal lines
   LINE (xmin, y)-(xmax, y), 8        '1.1x to get top line
NEXT y
'calculate the curve for different values of T and w
FOR T = 300 TO 380 STEP 80
   FOR w = 1 TO 25 STEP .01
      x = c2 / w / T                  'the expression for x
      M = c1 / w ^ 5 / (EXP(x) - 1)   'Planck's equation
      'PRINT w, M                     'option for printing
      PSET (w, M), 15                 'plotting the values;
   NEXT w                                     'the 12 gives a red line
NEXT T
'Calculate the Wien displacement law
FOR T = 100 TO 400 STEP .1
   wx = 2898.8 / T
```

```
    x = c2 / wx / T                     'the expression for x
    M - c1 / wx ^ 5 / (EXP(x) - 1)      'Planck's equation
    'PRINT w, M                         'option for printing
    PSET (wx, M), 15                    'plotting the values
NEXT T
RETURN

Frequencies:
CLS 0
SCREEN 9                                'set the screen for SVGA
VIEW (100, 60)-(600, 275)               'set an screen area
xmin = 0: xmax = 5000                   'set extreme ordinate values
ymin = 0: ymax = .0001                  'set extreme abscissa values
WINDOW (xmin, ymin)-(xmax, ymax)        'set appropriate window
'set up constants
c1 = 2 * pi * c * c * h                 'c1 in units of W/cm2
'locate titles and axis values appropriately
LOCATE 23, 15: PRINT "Figure 4-7. Spectral radiant exitance vs wavenumber"
LOCATE 22, 35: PRINT "Wavenumber [waves/cm]"
LOCATE 8, 1: PRINT "Spectral"
LOCATE 9, 1: PRINT "Radiant"
LOCATE 10, 1: PRINT "Exitance"
LOCATE 11, 1: PRINT "[uW/cm]"
LOCATE 20, 10: PRINT 0
LOCATE 17, 10: PRINT 5
LOCATE 14, 9: PRINT 10
LOCATE 11, 9: PRINT 15
LOCATE 8, 9: PRINT 20
LOCATE 5, 9: PRINT 25
LOCATE 21, 12: PRINT 0
LOCATE 21, 23: PRINT 1000
LOCATE 21, 35: PRINT 2000
LOCATE 21, 48: PRINT 3000
LOCATE 21, 61: PRINT 4000
LOCATE 21, 73: PRINT 5000
'draw the grid lines
FOR x = xmin TO xmax STEP 500           'vertical lines
  LINE (x, ymin)-(x, ymax), 8           'the 8 gives gray lines
NEXT x
FOR y = ymin TO 1.1 * ymax STEP .00001  'horizontal lines
  LINE (xmin, y)-(xmax, y), 8
NEXT y
LOCATE 16, 21: PRINT "300K"
LOCATE 9, 25: PRINT "350K"
'calculate the curve for different values of T and sigma
FOR T = 300 TO 350 STEP 50
  FOR sigma = 40 TO 5000 STEP 10
    x = c2 * sigma / 10000 / T          'the expression for x
    M = c1 * sigma ^ 3 / (EXP(x) - 1)   'Planck's equation
    'PRINT sigma, M                     'option for printing
    PSET (sigma, M), 15                 'plotting the values
  NEXT sigma
NEXT T
'Calculate the Wien displacement law
constant = 2.82143977# / 1.4399         'Wien constant for sigma
FOR T = 400 TO 100 STEP -1
  FOR sigma = 0 TO 5000 STEP 1000
    sigmamax = constant * T
    x = 1.4399 * sigmamax / T           'the expression for x in sigma
    M = c1 * sigmamax ^ 3 / (EXP(x) - 1)  'Planck's equation
```

```
     'PRINT sigmamax, M                    'option for printing
      PSET (sigmamax, M), 15               'plotting the values
   NEXT sigma
NEXT T
RETURN
SCREEN 9                              'set the screen for SVGA
VIEW (100, 40)-(600, 275)            'set an appropriate screen area
xmin = 0: xmax = 2                   'set extreme ordinate values
ymin = -3: ymax = 0                  'set extreme abscissa values
WINDOW (xmin, ymin)-(xmax, ymax)     'set appropriate calculational values
RETURN


Logwaves:
CLS 0                                     'clear the screen
'Figure 4. Log of spectral radiant exitance vs log of wavelength
CLS                                       'clear the screen
SCREEN 9                             'set the screen for EGA
VIEW (100, 40)-(600, 275)            'set an appropriate screen area
xmin = 0: xmax = 2                   'set extreme ordinate values
ymin = -3: ymax = 0                  'set extreme abscissa values
WINDOW (xmin, ymin)-(xmax, ymax)     'set appropriate calculational values
c1 = 2 * pi * c * c * h * 1E+16      'c1 in units of W/cm2/um
'locate titles and axis values appropriately
LOCATE 23, 15: PRINT "Figure 4.  Spectral radiant exitance vs Wavelength"
LOCATE 22, 35: PRINT "Wavelength [um]"
LOCATE 10, 1: PRINT "Spectral"
LOCATE 11, 1: PRINT "Radiant"
LOCATE 12, 1: PRINT "Exitance"
LOCATE 13, 1: PRINT "[mW/cm2/um]"
LOCATE 20, 8: PRINT .001
LOCATE 15, 9: PRINT .01
LOCATE 9, 10: PRINT .1
LOCATE 4, 10: PRINT 1
LOCATE 21, 11: PRINT 1
LOCATE 21, 43: PRINT 10
LOCATE 21, 74: PRINT 100
'draw the grid lines
FOR n = .001 TO .01 STEP .001
y = .4343 * LOG(n)
LINE (xmin, y)-(xmax, y), 8
NEXT n
FOR n = .01 TO .1 STEP .01
y = .4343 * LOG(n)
LINE (xmin, y)-(xmax, y), 8
NEXT n
FOR n = .1 TO 1 STEP .1
y = .4343 * LOG(n)
LINE (xmin, y)-(xmax, y), 8
NEXT n
FOR n = 1 TO 10 STEP 1
x = .4343 * LOG(n)
LINE (x, ymin)-(x, ymax), 8
NEXT n
FOR n = 10 TO 100 STEP 10
x = .4343 * LOG(n)
LINE (x, ymin)-(x, ymax), 8
NEXT n
LINE (xmin, ymin)-(xmax, ymin)
LINE (xmin, ymin)-(xmin, ymax)
LINE (xmax, ymin)-(xmax, ymax)
```

```
LINE (xmin, ymax)-(xmax, ymax)
LOCATE 8, 43: PRINT "900K"
LOCATE 19, 46: PRINT "300K"
'calculate the curve for different values of T and w
FOR T = 300 TO 900 STEP 600
  FOR w = 1 TO 20 STEP .01
    x = c2 / w / T 'the expression for x
    M = c1 / w ^ 5 / (EXP(x) - 1)          'Planck's equation
    'PRINT .4343 * LOG(w), .4343 * LOG(M)  'option for printing
    PSET (.4343 * LOG(w), .4343 * LOG(M)), 12  'plotting the values
  NEXT w
NEXT T
'Calculation of the Wien displacement law
FOR T = 1000 TO 200 STEP -1
    wx = 2898.8 / T
    x = c2 / wx / T 'the expression for x
    M = c1 / wx ^ 5 / (EXP(x) - 1)         'Planck's equation
    'PRINT w, M                            'option for printing
    PSET (.4343 * LOG(wx), .4343 * LOG(M)), 14 'plotting the values
NEXT T
RETURN

Lognumber:
CLS 0
SCREEN 9                            'set the screen for SVGA
VIEW (100, 60)-(600, 275)          'set an appropriate screen area
xmin = 1: xmax = 4                  'set extreme ordinate values
ymin = -6: ymax = -4               'set extreme abscissa values
WINDOW (xmin, ymin)-(xmax, ymax)   'set appropriate calculational values
'set up constants
c1 = 2 * pi * c * c * h
'locate titles and axis values appropriately
LOCATE 23, 15: PRINT "Figure 5.  Spectral radiant exitance vs wavenumber"
LOCATE 22, 35: PRINT "Wavenumber [waves/cm]"
LOCATE 8, 1: PRINT "Spectral"
LOCATE 9, 1: PRINT "Radiant"
LOCATE 10, 1: PRINT "Exitance"
LOCATE 11, 1: PRINT "[uW/cm]"
LOCATE 20, 11: PRINT 1
LOCATE 13, 10: PRINT 10
LOCATE 5, 9: PRINT 100
LOCATE 21, 12: PRINT 10
LOCATE 21, 32: PRINT 100
LOCATE 21, 52: PRINT 1000
LOCATE 21, 72: PRINT 10000
'draw the grid lines
FOR n = .00001 TO .0001 STEP .00001
y = .4343 * LOG(n)
LINE (xmin, y)-(xmax, y), 8
NEXT n
FOR n = .000001 TO .00001 STEP .000001
y = .4343 * LOG(n)
LINE (xmin, y)-(xmax, y), 8
NEXT n
FOR n = .1 TO 1 STEP .1
y = .4343 * LOG(n)
LINE (xmin, y)-(xmax, y), 8
NEXT n
FOR n = 10 TO 100 STEP 10
x = .4343 * LOG(n)
```

```
                         'The Radiation Signal to Noise Ratio--Figure 8-1
GOSUB Setup
GOSUB Calculate
GOSUB Plotit
again: GOTO again
END

Setup:
CLS : SCREEN 9
pi = 3.14159: c = 3E+10: h = 6.626E-34
c1 = 2 * pi * c ^ 2 * h * 1E+16: c2 = 14388
RETURN

Calculate:
FOR T = 250 TO 350 STEP 10
  w1 = 3: w2 = 5: dw = .01: GOSUB Integral
  ratio = LqdT / SQR(Lq)
'   PRINT T, USING "##.##^^^^   "; ratio, LdT,
  w1 = 8: w2 = 12: dw = .01: GOSUB Integral
  ratio = LqdT / SQR(Lq)
' print USING "##.##^^^^   "; ratio, LdT
NEXT T
RETURN

Plotit:
xmin = 250: xmax = 500: dx = 50: ymin = 6: ymax = 7: dy = 1
GOSUB Makegraph
FOR T = 250 TO 500 STEP .5
  w1 = 3: w2 = 5: dw = .01: GOSUB Integral
  ratio = LqdT / SQR(Lq)
  PSET (T, .4343 * LOG(ratio))
  w1 = 8: w2 = 12: dw = .01: GOSUB Integral
  ratio = LqdT / SQR(Lq)
  PSET (T, .4343 * LOG(ratio))
NEXT T

Integral:
dL = 0: L = 0: dLq = 0: Lq = 0: dLdT = 0: dLqdT = 0: LqdT = 0: LdT = 0
FOR w = w1 TO w2 STEP dw
  x = c2 / w / T
  photon = h * c * 10000! / w
  dL = c1 / pi / w ^ 5 / (EXP(x) - 1) * dw
  dLq = dL / photon
  dLdT = x * EXP(x) / (EXP(x) - 1) * dL / T
  dLqdT = x * EXP(x) / (EXP(x) - 1) * dLq / T
  L = L + dL
  Lq = Lq + dLq
  LqdT = LqdT + dLqdT
  LdT = LdT + dLdT
NEXT w
'PRINT T, w2, USING "##.##^^^^   "; L, Lq, LdT, LqdT
RETURN

Makegraph:
VIEW (100, 20)-(580, 270)
WINDOW (xmin, ymin)-(xmax, ymax)
GOSUB LabelGraph
FOR x = xmin TO xmax STEP dx
  LINE (x, ymin)-(x, ymax), 8
NEXT x
```

```
FOR i = 1 TO 10
  y1 - .4343 * LOG(i * 1000000!)
  LINE (xmin, y1)-(xmax, y1), 8
NEXT i
RETURN

LabelGraph:
LOCATE 2, 9: PRINT ymax
LOCATE 20, 9: PRINT ymin
LOCATE 21, 12: PRINT xmin
LOCATE 21, 72: PRINT xmax
LOCATE 22, 35: PRINT "Temperature [K]"
LOCATE 10, 5: PRINT "SNR"
LOCATE 23, 22: PRINT "Figure 8-1 The Radiation Signal to Noise Ratio"
RETURN
```

```
'Fresnel Equations for Reflectance and Transmittance--Fig. 10-1
GOSUB Setup
GOSUB Fresnel
GOSUB Transmission
GOSUB Index
END

Setup:
CLS
SCREEN 9
pi = 3.14159
RETURN

Fresnel:
CLS 0
VIEW (110, 20)-(600, 300), 1, 1
WINDOW (0, 0)-(90, 1)
FOR x = 0 TO 90 STEP 10: LINE (x, 0)-(x, 1): NEXT x
FOR y = 0 TO 1 STEP .1: LINE (0, y)-(90, y): NEXT y
LOCATE 9, 1: PRINT "Transmittance"
LOCATE 11, 5: PRINT "or"
LOCATE 13, 1: PRINT "Reflectance"
LOCATE 4, 55: PRINT "Reflectance"
LOCATE 6, 16: PRINT "Transmittance"
LOCATE 23, 33: PRINT "Angle of Incidence"
LOCATE 2, 10: PRINT "1.0"
LOCATE 4, 10: PRINT "0.9"
LOCATE 6, 10: PRINT "0.8"
LOCATE 8, 10: PRINT "0.7"
LOCATE 10, 10: PRINT "0.6"
LOCATE 12, 10: PRINT "0.5"
LOCATE 14, 10: PRINT "0.4"
LOCATE 16, 10: PRINT "0.3"
LOCATE 18, 10: PRINT "0.2"
LOCATE 20, 10: PRINT "0.1"
LOCATE 22, 10: PRINT "0.0"
LOCATE 23, 14: PRINT "0"
LOCATE 23, 74: PRINT "90"
  FOR n = 1.5 TO 1.5
    FOR angle = .1 TO 90 STEP .1
      theta = angle * pi / 180: a = SIN(theta) / n
       theta1 = ATN(a / SQR(1 - a ^ 2))
         Rs = (SIN(theta - theta1) / SIN(theta + theta1)) ^ 2
          Rp = (TAN(theta - theta1) / TAN(theta + theta1)) ^ 2
           Ts = (2 * SIN(theta1) * COS(theta) / SIN(theta + theta1)) ^ 2
            Tp = (2 * SIN(theta1) * COS(theta) / SIN(theta + theta1) * COS(theta -
             PSET (angle, Rs), 14: PSET (angle, Rp), 14: PSET (angle, Ts), 12: PSET
              NEXT angle
                NEXT n

DO
LOOP WHILE INKEY$ = ""
RETURN

Transmission:
CLS 0
VIEW (110, 20)-(600, 300), 1, 1
WINDOW (0, 0)-(1, 1)
LINE (0, 0)-(0, 1)
LINE (0, 0)-(1, 0)
LOCATE 23, 40: PRINT "Single Surface Reflectivity"
```

```
LOCATE 2, 11: PRINT "1.0"
LOCATE 4, 11: PRINT "0.9"
LOCATE 6, 11: PRINT "0.8"
LOCATE 8, 11: PRINT "0.7"
LOCATE 10, 11: PRINT "0.6"
LOCATE 12, 11: PRINT "0.5"
LOCATE 14, 11: PRINT "0.4"
LOCATE 16, 11: PRINT "0.3"
LOCATE 18, 11: PRINT "0.2"
LOCATE 20, 11: PRINT "0.1"
LOCATE 22, 11: PRINT "0.0"
FOR x = 0 TO 1 STEP .1: LINE (x, 0)-(x, 1), 14: NEXT x
FOR y = 0 TO 1 STEP .1: LINE (0, y)-(1, y), 14: NEXT y
FOR t = 0 TO 1.1 STEP .1
 FOR r = 0 TO 1 STEP .001
  tau = (1 - r) ^ 2 * t / (1 - r ^ 2 * t ^ 2)
   PSET (r, tau), 14
    NEXT r
     NEXT t
LOCATE 17, 55: PRINT "Transmission"
DO
LOOP WHILE INKEY$ = ""
FOR t = 0 TO 1.1 STEP .1
 FOR r = 0 TO 1 STEP .001
  rho = r + (1 - r) ^ 2 * r * t ^ 2 / (1 - r ^ 2 * t ^ 2)
   PSET (r, rho), 12
    NEXT r
     NEXT t
LOCATE 3, 55: PRINT "Reflection"
DO
LOOP WHILE INKEY$ = ""
DO
LOOP WHILE INKEY$ = ""
RETURN

Index:
CLS 0
VIEW (110, 20)-(600, 300), 1, 1
WINDOW (0, 0)-(6, 1)
LINE (0, 0)-(0, 1)
LINE (0, 0)-(6, 0)
LOCATE 11, 1: PRINT "Transmittance"
LOCATE 23, 40: PRINT "Refractive Index"
LOCATE 2, 11: PRINT "1.0"
LOCATE 4, 11: PRINT "0.9"
LOCATE 6, 11: PRINT "0.8"
LOCATE 8, 11: PRINT "0.7"
LOCATE 10, 11: PRINT "0.6"
LOCATE 12, 11: PRINT "0.5"
LOCATE 14, 11: PRINT "0.4"
LOCATE 16, 11: PRINT "0.3"
LOCATE 18, 11: PRINT "0.2"
LOCATE 20, 11: PRINT "0.1"
LOCATE 22, 11: PRINT "0.0"
FOR x = 0 TO 6: LINE (x, 0)-(x, 1), 14: NEXT x
FOR y = 0 TO 1 STEP .1: LINE (0, y)-(6, y), 14: NEXT y
FOR n = 0 TO 6 STEP .001: y = 2 * n / (n * n + 1): PSET (n, y), 12: NEXT n
DO
LOOP WHILE INKEY$ = ""

RETURN
```

```
                    'Multiple reflection, reflection and transmission--Fig 10-2
GOSUB Setup
GOSUB Plotit
Again: GOTO Again
END

Setup:
CLS : SCREEN 9
pi = 3.14159
RETURN

Plotit:
xmin = 0: xmax = 1: dx = .1: ymin = 0: ymax = 1: dy = .1
GOSUB MakeGraph
GOSUB LabelGraph
  FOR tau = 1 TO 0 STEP -.1
    FOR rho = 0 TO 1 STEP .001
      taueff = (1 - rho) ^ 2 * tau / (1 - rho ^ 2 * tau ^ 2)
      rhoeff = rho + (1 - rho) ^ 2 * tau ^ 2 / (1 - rho ^ 2 * tau ^ 2)
      PSET (rho, taueff)
      PSET (rho, rhoeff)
      NEXT rho
  NEXT tau
RETURN

MakeGraph:
VIEW (100, 20)-(580, 270), , 15
WINDOW (xmin, ymin)-(xmax, ymax)
FOR x = xmin TO xmax STEP dx
  LINE (x, ymin)-(x, ymax), 8
NEXT x
FOR y = ymin TO 2 * ymax STEP dy
  LINE (xmin, y)-(xmax, y), 8
NEXT y
RETURN

LabelGraph:
LOCATE 2, 9: PRINT ymax
LOCATE 20, 9: PRINT ymin
LOCATE 21, 12: PRINT xmin
LOCATE 21, 72: PRINT xmax
LOCATE 21, 32: PRINT "Single-surface reflectivity"
LOCATE 22, 11: PRINT "Figure 10-2 Transmission and Reflection "
LOCATE 8, 55: PRINT "Transmission"
LOCATE 16, 55: PRINT "Reflection"
LOCATE 3, 55: PRINT "tau = 1"
LOCATE 7, 59: PRINT "tau = 0"
LOCATE 19, 20: PRINT "tau = 0"
LOCATE 13, 42: PRINT "tau = 1"
RETURN
```

```
                        'Transmittance as a Function of Refractive Index--Figure 10-3
GOSUB Setup
GOSUB MakeGraph
GOSUB MakePlot
again: GOTO again
END

Setup:
CLS
SCREEN 9
pi = 3.14159
xmin = 1: xmax = 6: ymin = 0: ymax = 1
RETURN

MakeGraph:
VIEW (100, 10)-(580, 270), , 15
WINDOW (xmin, ymin)-(xmax, ymax)
FOR x = xmin TO xmax STEP 1
  LINE (x, ymin)-(x, ymax), 8
NEXT x
FOR y = ymin TO ymax STEP .1
  LINE (xmin, y)-(xmax, y), 8
NEXT y
LOCATE 1, 10: PRINT ymax
LOCATE 20, 10: PRINT ymin
LOCATE 21, 12: PRINT xmin
LOCATE 21, 72: PRINT xmax
LOCATE 22, 20: PRINT "Figure 10-3.  Transmittance vs Refractive Index"
RETURN

MakePlot:
FOR n = 1 TO 6 STEP .01
  tau = 2 * n / (n ^ 2 + 1)
  PSET (n, tau)
NEXT n
RETURN
```

Appendix B: EXAMPLE PROGRAMS

Chapter 14 pulls together many of the concepts by carrying out the beginning designs for several different applications. This appendix provides the programs that were used to obtain the calculations and the outputs of those programs.

B.1 The Programs

The first example is the night-driving system. There are two programs for this, one for photon detectors and one for thermal detectors. The first program is for the photonic night driver. The standard format I like to use is to provide all the input values in the **Setup** subroutine. CLS and SCREEN clear the screen and set the monitor for SVGA in case I want to plot. For some reason this BASIC does not know the value of π. So I give it and the atomic constants that will get the first and second radiation constants for the Planck equation. Note that c_1 is in terms of c and h. A multiplier of 10^{16} is necessary to get it in terms of centimeters and micrometers. Then I set the efficiencies and the transmissions. I use the last two lines alternately. If the last line is REMARKED, then the system works in the LWIR; if not, in the MWIR. I think the geometry and BANDWIDTH subroutines are clear. There are a lot of printing instructions to make sure things are the way they should be as the program proceeds. The SENSITIVITY subroutine uses the INTEGRATION subroutine that calculates all the blackbody functions. This same subroutine can be used for all the programs. Then there is a little optics calculation using the scheming equations.

The second program is for ambient-temperature, thermal detectors. The inputs are a little different. The specific detectivity is given and so are some of the properties of the optics. Only the one spectral band is used. Most everything else is the same.

The same scheme is used for ICBMs, but now the required values for out-of-field rejection and system temperature are determined from solar and terrestrial irradiances on the aperture. Then an iteration is used to get the temperature at which the system must be operated.

The PEACEKEEPER program uses a different method for calculating the bandwidth, since it is a strip-mapper, but it follows the textural prescription. Great similarities will be found in the other subroutines and this is no coincidence.

B.2 The Strategy

There are different ways to set up programs like this. I have chosen to write several different programs for different types of devices. They could be combined, and all use the common integral and, maybe, geometry and bandwidth subroutines. QuickBASIC makes it easy to Copy and Paste. So I did it this way. Then I don't have to make many selections at run time.

It should also be clear that I could have asked for input data on an interactive basis, but since these are my programs and not used by anyone else, I chose to do it with a series of equalities. As I see it, there are two major problems with the other method: The first is that there is a lot of selecting to do at run time. The second is that in writing such a program, the author has to think of just about every possibility, and I have trouble remembering why I came to the living room! These programs have all been done in an early version of QuickBASIC, Version 4.0, but the programming is consonant with later versions, with QBASIC and with Visual BASIC. I hope they are useful.

```
'Night Driver -- Photon Detectors
LPRINT
LPRINT , , "Photon Night Driver"
GOSUB Setup
GOSUB GEOMETRY
GOSUB BANDWIDTH
GOSUB SENSITIVITY
GOSUB Optics
END

Setup:
CLS : SCREEN 9
pi = 3.14159: c = 3E+10: h = 6.626E-34
c1 = 2 * pi * c ^ 2 * h * 1E+16: c2 = 14388
eta = .8: etasc = 1: etaff = .5: etacs = 1
tauatm = .5: tauopt = .5: Ddet = .0025
T = 300: w1 = 8: w2 = 12: dw = .01
T = 300: w1 = 3: w2 = 5: dw = .01: eta = .01
RETURN

GEOMETRY:
reselm = 5            'cm
halffield = 15           'degrees
distance = 10000     'cm
angle = reselm / distance
Npx = 2 * halffield / 180 * pi / angle
Ddif = 1.22 * w2 / 10000 / angle         'watch the units
LPRINT , , "GEOMETRY"
LPRINT " reselm", "distance", " angle", "Ddif", "Count"
LPRINT " [cm]", "   [cm]", "[rad]", "[cm]", "  [--]"
LPRINT reselm, distance, USING "####.####       "; angle, Ddif, Npx
RETURN

BANDWIDTH:
LPRINT
LPRINT , , "BANDWIDTH"
td = 1 / 30
B = 1 / 2 / td
LPRINT , , "Bandwidth"
LPRINT , , USING "##.##^^^^      "; B
LPRINT
RETURN

SENSITIVITY:
factor = pi / 4 * tauatm * tauopt * eta * etasc * etaff * etacs
GOSUB Integral
SNRDT = factor * w2 / 10000 * LqdT / SQR(Lq) / SQR(B)
LPRINT
NETD = 1 / SNRDT
LPRINT "w1", "w2", "NETD"
LPRINT "um", "um", "[K]"
LPRINT w1, w2, USING "##.###"; NETD
LPRINT
RETURN

Optics:
efl = Ddet / angle
F# = efl / Ddif
Bdif = 2.44 * w2 / 10000 / Ddif
Bsa = 1 / 128 / F# ^ 3
```

```
Bca = halffield / 180 * pi / 16 / F# ^ 2
Baa = (halffield / 180 * pi) ^ 2 / 2 / F#
LPRINT , , "OPTICS"
LPRINT "Dopt", "Ddet", "angle", "focalength", "F/#"
LPRINT Ddif, Ddet, angle, USING "##.##        "; efl, F#
LPRINT
LPRINT "Angle", "Bdif", "Bsa", "Bca", "Baa"
LPRINT "[rad]", "[rad]", "[rad]", "[rad]", "[rad]"
LPRINT USING "##.####      "; angle, Bdif, Bsa, Bca, Baa
RETURN



Integral:
dL = 0: L = 0: dLq = 0: Lq = 0: dLdT = 0: dLqdT = 0: LqdT = 0: LdT = 0
FOR w = w1 TO w2 STEP dw
  x = c2 / w / T
  photon = h * c * 10000! / w
  dL = c1 / pi / w ^ 5 / (EXP(x) - 1) * dw
  dLdT = x * EXP(x) / (EXP(x) - 1) * dL
  dLq = dL / photon
  dLqdT = x * EXP(x) / (EXP(x) - 1) * dLq
  L = dL + L
  LdT = dLdT + LdT
  Lq = Lq + dLq
  LqdT = LqdT + dLqdT
NEXT w
LdT = LdT / T: LqdT = LqdT / T
RETURN
```

```
                           Photon Night Driver Output
                           GEOMETRY
    reselm        distance    angle      Ddif           Count
    [cm]            [cm]     [rad]        [cm]           [--]
    5             10000         0.0005         1.2200        1047.1967

                           BANDWIDTH
                           Bandwidth
                           1.50E+01


    w1            w2          NETD
    um            um          [K]
    3             5           3.407

                           OPTICS
    Dopt          Ddet        angle      focalength     F/#
    1.22          .0025       .0005      5.00           4.10

    Angle         Bdif        Bsa        Bca            Baa
    [rad]         [rad]       [rad]      [rad]          [rad]
    0.0005        0.0010      0.0001     0.0010         0.0084

                           Photon Night Driver Output
                           GEOMETRY
    reselm        distance    angle      Ddif           Count
    [cm]            [cm]     [rad]        [cm]           [--]
    5             10000         0.0005         2.9280        1047.1967

                           BANDWIDTH
                           Bandwidth
                           1.50E+01


    w1            w2          NETD
    um            um          [K]
    8             12          0.006

                           OPTICS
    Dopt          Ddet        angle      focalength     F/#
    2.928         .0025       .0005      5.00           1.71

    Angle         Bdif        Bsa        Bca            Baa
    [rad]         [rad]       [rad]      [rad]          [rad]
    0.0005        0.0010      0.0016     0.0056         0.0201
```

```
'Night Driver--Ambient Temperature Detectors
LPRINT
LPRINT , , "Ambient-Temperature Night Driver"
GOSUB Setup
GOSUB GEOMETRY
GOSUB BANDWIDTH
GOSUB Thermals
END

Setup:
CLS : SCREEN 9
pi = 3.14159: c = 3E+10: h = 6.626E-34
c1 = 2 * pi * c ^ 2 * h * 1E+16: c2 = 14388
tauatm = .5: tauopt = .5
Dstar = 1E+09
Dopt = 3: Ddet = .005
T = 300: w1 = 8: w2 = 12: dw = .01
RETURN

GEOMETRY:
reselm = 10              'cm
halffield = 15           'degrees
distance = 10000         'cm
angle = reselm / distance
Npx = halffield / 180 * pi / angle
Ddif = 1.22 * w2 / 10000 / angle
LPRINT , , "GEOMETRY"
LPRINT " reselm", "distance", " angle", "Ddif", "Count"
LPRINT " [cm]", "   [cm]", "[rad]", "[cm]", "  [--]"
LPRINT reselm, distance, USING "##.##^^^^     "; angle, Dopt, Npx
RETURN

BANDWIDTH:
LPRINT
LPRINT , , "BANDWIDTH"
td = 1 / 30
B = 1 / 2 / td
LPRINT , , "Bandwidth"
LPRINT , , USING "##.##^^^^     "; B
LPRINT
RETURN

Integral:
dL = 0: L = 0: dLq = 0: Lq = 0: dLdT = 0: dLqdT = 0: LqdT = 0: LdT = 0
FOR w = w1 TO w2 STEP dw
  x = c2 / w / T
  photon = h * c * 10000! / w
  dL = c1 / pi / w ^ 5 / (EXP(x) - 1) * dw
  dLdT = x * EXP(x) / (EXP(x) - 1) * dL
  dLq = dL / photon
  dLqdT = x * EXP(x) / (EXP(x) - 1) * dLq
  L = dL + L
  LdT = dLdT + LdT
  Lq = Lq + dLq
  LqdT = LqdT + dLqdT
NEXT w
LdT = LdT / T: LqdT = LqdT / T
RETURN

Thermals:
```

```
focalength = Ddet / angle: F# = focalength / Dopt
LPRINT , , "SENSITIVITY"
LPRINT "Dopt", "Ddet", "angle", "focalength", "F/#"
LPRINT Dopt, Ddet, angle, USING "##.##          "; focalength, F#
factor = pi / 4 * tauatm * tauopt * Dopt / F# * angle * Dstar / SQR(B)
GOSUB Integral
SNRTD = factor * LdT
NETD = 1 / SNRTD
LPRINT
LPRINT , , "NETD"
LPRINT , , USING "##.##"; NETD
RETURN
```

Ambient-Temperature Night Driver Output

		GEOMETRY		
reselm	distance	angle	Ddif	Count
[cm]	[cm]	[rad]	[cm]	[--]
10	10000	1.00E-03	3.00E+00	2.62E+02

BANDWIDTH
Bandwidth
1.50E+01

SENSITIVITY
Dopt	Ddet	angle	focalength	F/#
3	.005	.001	5.00	1.67

NETD
0.17

```
'ICBM
LPRINT , , "ICBM"
LPRINT
LPRINT
GOSUB Setup
GOSUB Geometry
GOSUB Bandwidth
GOSUB Sensitivity
GOSUB Solar
GOSUB Earth
GOSUB SelfEmission
END

Setup:
CLS : SCREEN 9
pi = 3.14159: c = 3E+10: h = 6.626E-34
c1 = 2 * pi * c ^ 2 * h * 1E+16: c2 = 14388
F# = 3: tf = 1: fullfield = .2
eta = .8: tauatm = 1: tauopt = .1: R = 5000000!: g = 4
epsA = 1
res = 1000: m = 1000: SNR = 10: w1 = 2: w2 = 24: dw = .01
LPRINT , , "INPUTS"
LPRINT
LPRINT "Range", "Resolution", "Frame", "epsA", "Field"
LPRINT , , "Time", , "square"
LPRINT "[m]", "[m]", "[sec]", "[sq m]", "[rad]"
LPRINT R, res, tf, epsA, fullfield
RETURN

Geometry:
angle = res / R
Npx = fullfield / angle
Dopt = 2.44 * w2 / 10000! / angle
Aopt = pi / 4 * Dopt ^ 2:
focalength = Dopt * F#
Ddet = angle * focalength
Adet = Ddet ^ 2
LPRINT
LPRINT , , " GEOMETRY"
LPRINT "Angle", "Field", "Pixels", "Elements", "g"
LPRINT angle, fullfield, Npx, m, g
LPRINT
LPRINT "F#", "f", "Ddet", "Dopt"
LPRINT "[--]", "[cm]", "[cm]", "[cm]"
LPRINT F#, focalength, Ddet, Dopt
LPRINT
LPRINT "Optics ", "Detector", "QE", "tauatm", "tauopt"
LPRINT "Area", "Area"
LPRINT "[sq cm]", "[sq cm]", "[--] ", "[--]", "[--]"
LPRINT USING "####        "; Aopt,
LPRINT USING "##.##^^^^        "; Adet,
LPRINT eta, tauatm, tauopt
RETURN

Bandwidth:
td = tf / Npx: B = 1 / 2 / td:
LPRINT
LPRINT , , "Bandwidth"
LPRINT , , "[Hz]"
LPRINT , , B; ""
```

```
RETURN

Sensitivity:
eff = SQR(eta) * tauatm * tauopt / SQR(g)
geo = Aopt * epsA / SQR(Adet) / R ^ 2
overallfactor = eff * geo / SQR(B)
LPRINT
LPRINT , , "SENSITIVITY"
LPRINT , "SNR", "  Signal", "Eq Required"
T = 250
GOSUB Integral
EqReq = (overallfactor * Lq / SNR) ^ 2
LPRINT , SNR, USING "##.##^^^^    "; overallfactor * Lq, EqReq
RETURN

Integral:
dL = 0: L = 0: dLq = 0: Lq = 0: dLdT = 0: dLqdT = 0: LqdT = 0: LdT = 0
FOR w = w1 TO w2 STEP dw
  x = c2 / w / T
  photon = h * c * 10000! / w
  dL = c1 / pi / w ^ 5 / (EXP(x) - 1) * dw
  dLq = dL / photon
  Lq = Lq + dLq
NEXT w
RETURN

Solar:
T = 5900
GOSUB Integral
Omega = pi / 4 * (33 / 60 * pi / 180) ^ 2
Eq = Lq * Omega
LPRINT
LPRINT , , "RATIOS"
LPRINT "EqSolar/EqReq ", " EqEarth/EqReq ", "System Temperature"
LPRINT "[--]", "[--]", "[K]"
LPRINT USING "##.##^^^^    "; Eq / EqReq,
RETURN

Earth:
T = 275
GOSUB Integral
Omega = pi
Eq = Lq * Omega
LPRINT , , USING "##.##^^^^    "; Eq / EqReq,
RETURN

SelfEmission:
rho = .95: F = 3
Omega = pi / 4 / F ^ 2
FOR T = 50 TO 40 STEP -1
  w1 = 5: w2 = 24: dw = .01: GOSUB Integral
  Eq = (1 - rho) * (1 + rho + rho ^ 2) * Omega * Lq
  IF Eq < EqReq THEN LPRINT , , T: END
NEXT T
```

ICBM Output

INPUTS

Range	Resolution	Frame Time	epsA	Field square
[m]	[m]	[sec]	[sq m]	[rad]
5000000	1000	1	1	.2

GEOMETRY

Angle	Field	Pixels	Elements	g
.0002	.2	1000	1000	4

F#	f	Ddet	Dopt	
[--]	[cm]	[cm]	[cm]	
3	87.84001	.017568	29.28	

Optics Area	Detector Area	QE	tauatm	tauopt
[sq cm]	[sq cm]	[--]	[--]	[--]
673	3.09E-04	.8	1	.1

Bandwidth
[Hz]
 500

SENSITIVITY

	SNR	Signal	Eq Required	
	10	1.14E+06	1.29E+10	

RATIOS

EqSolar/EqReq		EqEarth/EqReq		System Temperature
[--]	[--]	[K]		
1.12E+07		1.33E+08		46

```
                        'Peacekeeper
GOSUB Setup
GOSUB GEOMETRY
GOSUB BANDWIDTH
GOSUB Sensitivity
GOSUB Optics
END

Setup:
CLS : SCREEN 9
pi = 3.14159: c = 3E+10: h = 6.626E-34
c1 = 2 * pi * c ^ 2 * h * 1E+16: c2 = 14388
m = 10000
eta = .8: tauatm = .5: tauopt = .5:  g = 4
etasc = .5: etacs = 1
Aopt = pi / 4 * Dopt ^ 2:  Ddet = .005
height = 10000000 'cm
v = 700000'cm/s
T = 300: w1 = 8: w2 = 12: dw = .01
RETURN

GEOMETRY:
gsd = 1000
halffield = 45
angle = gsd / height * (COS(halffield)) ^ 2
gsdnadir = height * angle
Dopt = 1.22 * w2 / 10000 / angle
PRINT , , "GEOMETRY"
PRINT " gsd", "height", "gsdnadir ", " angle", "Dopt"
PRINT " [cm]", "   [cm]", "  [cm]", "[cm]", "[cm"
PRINT gsd, height, USING "##.##^^^^      "; gsdnadir, angle, Dopt
PRINT
RETURN

BANDWIDTH:
linetime = gsdnadir / v
Npx = 2 * halffield / angle
td = linetime / Npx * m
B = 1 / 2 / td
PRINT , , " BANDWIDTH"
PRINT " line time", "   Npx", "dwell time", "bandwidth"
PRINT "   [sec]", "  [--]", "   [sec]", "  [Hz]"
PRINT USING "##.##^^^^      "; linetime, Npx, td, B
PRINT
RETURN

Sensitivity:
PRINT , , "SENSITIVITY"
efficiency = SQR(eta) * tauatm * tauopt * SQR(etasc) * SQR(etacs) / SQR(g)
PRINT "eta", "tauatm", "tauopt", " g", "etasc"
PRINT eta, tauatm, tauopt, g, etasc
PRINT
GOSUB Integral
PRINT "efficiency", "dLq/dT", "sqr(Lq)", "w dLq/dT/sqr(Lq)"
PRINT USING "##.##^^^^      "; efficiency, LqdT, SQR(Lq), w * .0001 * LqdT / SQR(L
SNRTD = efficiency * w * .0001 * LqdT / SQR(Lq) / SQR(B)
NETD = 1 / SNRTD
PRINT
PRINT , "  SNRTD", , "NETD"
PRINT , USING "##.##^^^^                "; SNRTD, NETD
```

```
RETURN

Integral:
dL = 0: L = 0: dLq = 0: Lq = 0: dLdT = 0: dLqdT = 0: LqdT = 0: LdT = 0
FOR w = w1 TO w2 STEP dw
  x = c2 / w / T
  photon = h * c * 10000! / w
  dL = c1 / pi / w ^ 5 / (EXP(x) - 1) * dw
  dLq = dL / photon
  dLqdT = x * EXP(x) / (EXP(x) - 1) * dLq
  L = dL + L
  Lq = Lq + dLq
  LqdT = LqdT + dLqdT
NEXT w
LdT = LdT / T: LqdT = LqdT / T
RETURN

Optics:
PRINT
PRINT , , "OPTICS"
focalength = Ddet / angle
F# = focalength / Dopt
PRINT "Ddet", "f", "Dopt", "F#"
PRINT "[cm]", "[cm]", "[cm]", "[--]"
PRINT Ddet, USING "###.##     "; focalength, Dopt, F#
Bdiff = 1.22 * w2 / 10000 / Dopt
Bsa = 1 / 128 / F# ^ 3
Bca = halffield / 180 * pi / 16 / F# ^ 2
Baa = halffield / 180 * pi ^ 2 / 2 / F#
PRINT
PRINT " angle", "Bdiff", "Bsa", "Bca", "Baa"
PRINT "[rad]", "[rad]", "[rad]", "[rad]", "[rad]"
PRINT USING "##.##^^^^     "; angle, Bdiff, Bsa, Bca, Baa
```

PEACEKEEPER OUTPUT

GEOMETRY

gsd [cm] 1000	height [cm] 1E+07	gsdnadir [cm] 2.76E+02	angle [cm] 2.76E-05	Dopt [cm 5.31E+01

BANDWIDTH

line time [sec] 3.94E-04	Npx [--] 3.26E+06	dwell time [sec] 1.21E-06	bandwidth [Hz] 4.14E+05

SENSITIVITY

eta .8	tauatm .5	tauopt .5	g 4	etasc .5

efficiency 7.91E-02	dLq/dT 3.13E+15	sqr(Lq) 4.40E+08	w dLq/dT/sqr(Lq) 8.53E+03

SNRTD 1.05E+00		NETD 9.54E-01

OPTICS

Ddet [cm] .005	f [cm] 181.18	Dopt [cm] 53.05 3.42	F# [--]

angle [rad] 2.76E-05	Bdiff [rad] 2.76E-05	Bsa [rad] 1.96E-04	Bca [rad] 4.21E-03	Baa [rad] 3.61E-01

BIBLIOGRAPHY

This list consists of all the books I know on infrared, radiometry, and detectors. Some I have even loved. Because most were published some time ago, they do not cover some of the modern techniques that use arrays and computer techniques. Many are also out of print, and require a good bookseller (or a good library) to obtain.

Hudson[1] is still probably the best all-around treatment of infrared system design, even though it was written many years ago. Unfortunately it does not deal with arrays and real-time imagers. The book arose from an educational program he ran at Hughes and therefore has much of input from the Hughes infrared engineers. It is distinguished by a fine compendium of applications, many from the patent literature.

Lloyd[2] is still the best book on real-time imagers; it arose from his experiences at the Army Night Vision Lab and the Honeywell Radiation Center. It includes good information on the properties of the eye, MRTDs, and similar figures of merit, and a nice treatment of the transfer-function approach to design.

The next three books all were published at just about the same time. They are a little older than the Hudson volume, but provide good information on both components and design approaches. The text by Kruse, McGlauchlin, and McQuistan[3] at the Honeywell Research Center has the best treatment of detectors; that by Jamieson, McFee, Plass, Grube, and Richards[4] at Aerojet deals more with detection probabilities and system analysis; that by Holter, Suits, Nudelman, Wolfe, and Zissis[5] at the University of Michigan, dubbed the purple peril, is the most balanced.

One of the earliest books on infrared systems, and still a good one, is the treatise by Smith, Jones, and Chasmar.[6]

Seyrafi[7] is a more general treatment of both infrared and electro-optics. It was printed by his own company and therefore did not have the benefit of independent review. It is good.

Ghys[8] and Wallace[9] are discussions of medical thermography, largely from the clinical point of view. Unfortunately for most, Ghys is in French, Canadian French.

Two new additions are Spiro and Schlesinger[10] and Holst.[11] Spiro and Schlesinger both worked at Aerospace, and the book has a resultant spacey flavor that incorporates considerations of staring space systems and their comparisons with scanners. The book by Holst on testing is devoted mostly to testing of real-time imagers, including MRTDs and others figures of merit and performance. It too was published by his own company, but was refereed by several independent authorities.

Four modest books are the ones by Morten et al.,[12] Kemp,[13] Burnay, [14] and Vanzetti.[15] The first of these is a compilation of applications by Mullard employees, published by Mullard, perhaps to help sell detectors. The second is one of the once-famous Sams series for technicians. The third is a compendium of chapters by different authors on a variety of thermal-imaging applications. The fourth is a collection of nondestructive testing applications. Each of these makes an interesting evening's reading.

Two good books on laboratory techniques are those by Conn and Avery[16] and Simon.[17]

A good book, albeit in French, is the one by Hadni.[18] It deals with the long wave infrared in hundreds of micrometers, tenths of millimeters.

Arams,[19] Hudson and Hudson,[20] and Johnson and Wolfe[21] are all collections of significant papers. The first two concentrate on detectors; the third is a more comprehensive treatment of infrared technology. The papers in all of them are historical yet still useful.

The best all-around book on detectors is by Dereniak and Crowe.[22] The two by Willardson and Beer[23] are collections of articles by authorities on specific detectors. There is more solid-state physics in these for some system designers.

A fine book on the electronics for and testing of infrared detectors has been written relatively recently by Vincent.[24]

The best overall book on radiometry is by Grumm and Becherer.[25] Wyatt[26] has written two. They both show his hands-on familiarity with the subject. He is the one author who embraced the Nicodemus system of nomenclature. Boyd's book[27] is probably just right for his course at Rochester, but it falls short of discussing calibration, standards, and normalization. Hengtsberger[28] wrote a detailed account on the use of radiometry based on electrical equivalency. It is specialized and is good in its specialty.

There are three handbooks pertinent to infrared technology. The first was compiled by Wolfe[29] in 1969. It was completely revised in 1984 by Wolfe and Zissis.[30] A still newer version, edited by Accetta and Shumaker,[31] covers electro-optics and a variety of applications as well. Although the IR and EO Handbook covers some of the same material as the previous versions, it treats detectors, for instance, with greater depth and includes the applications. The Optical Society of America has recently published a two-volume handbook of optics[32] that has several pertinent chapters.

There are surely more to come, and there are other books that are pertinent—some, for instance, on optical design, solid-state physics, and electronics. There are many interesting patents, especially on scanning systems, and there have been a few special issues of journals on infrared. These include two in *Applied Optics* and two more recent ones in *Optical Engineering* and one in the Proceedings of the Institute of Radio Engineers[33] (now the IEEE). One of the best of these is a critical review by Smith.[34] There is also a journal, *Infrared Physics and Technology,* devoted solely to infrared developments.

1.   R. D. Hudson, *Infrared System Engineering,* Wiley, 1969. I understand this can be obtained from OpAmp Technical Books, 800/468-4322, for over $100.

2.   J. M. Lloyd, *Thermal Imaging Systems,* Plenum Press, 1975

3.   P. W. Kruse, L. D. McGlauchlin, and R. R. McQuistan, *Elements of Infrared Technology,* Wiley, 1962.

4.   J. A. Jamieson, R. H. McFee, G. N. Plass, R. H. Grube, and R. G. Richards, *Infrared Physics and Engineering,* McGraw-Hill, 1963.

5. M. R. Holter, S. Nudelman, G. H. Suits, W. L. Wolfe, and G. J. Zissis, *Fundamentals of Infrared Technology,* Macmillan, 1962.

6. R. A. Smith, F. E. Jones, R. P. Chasmar, *The Detection and Measurement of Infrared Radiation,* Clarendon Press, 1968.

7. K. Seyrafi, *Electro-Optical Systems Analysis,* Electro-Optical Research Company, 1973.

8. R. Ghys, *Thermographie Medicale,* Somabed, Ltee, 1973.

9. J. D. Wallace and C. M. Cade, *Clinical Thermographie,* CRC Press, 1975.

10. I. J. Spiro and M. Schlesinger, *Infrared Technology Fundamentals,* Marcel Dekker, 1989.

11. G. C. Holst, *Testing and Evaluation of Infrared Imaging Systems,* JCD Publishing Company and SPIE Press, 1995.

12. F. D. Morten, T. J. Jarratt, P. R. D. Coleby, R. A. Lockett, M. H. Jervis, and R. J. Hutchinson, *Applications of Infrared Detectors,* Mullard, 1971.

13. B. Kemp, *Infrared,* Sams, 1972.

14. S. G. Burnay, T. L. Williams, C. H. Jones, *Applications of Thermal Imaging,* Adam Hilger, 1988.

15. R. Vanzetti. *Practical Applications of Infrared Techniques,* Wiley, 1972.

16. G. K. T. Conn and D. G. Avery, *Infrared Methods,* Academic Press, 1960.

17. I. Simon, *Infrared Radiation,* Van Nostrand, 1966.

18. A. Hadni, *L'infrarouge lointain,* Presses Universitaires de France, 1969.

19. F. R. Arams, *Infrared and Millimeter Wave Detectors,* Artech House, 1973.

20. R. D. Hudson and J. W. Hudson, *Infrared Detectors,* Dowden, Hutchison and Ross, 1975.

21. R. B. Johnson and W. L. Wolfe, *Selected Papers on Infrared Design,* SPIE Press, 1985.

22. E. L. Dereniak and D. Crowe, *Optical Radiation Detectors,* Wiley, 1984.

23. R. Willardson and R. Beer, Semiconductors and Semimetals, *Infrared Detectors,* **5,** 1970 and **12,** 1977.

24. D. Vincent, *Fundamentals of Infrared Detector Operation and Testing,* Wiley, 1989.

25. F. Grumm and R. J. Becherer, *Radiometry, Optical Radiation Measurements,* **1,** Academic Press, 1979.

26. C. Wyatt, *Radiometry,* Macmillan, 1975; *Radiometric System Design,* Macmillan 1987.

27. R. Boyd, *Radiometry and the Detection of Optical Radiation,* Wiley, 1982.

28. H. Hengstberger, *Absolute Radiometry,* Academic Press, 1989.

29. W. L. Wolfe, ed., *Handbook of Military Infrared Technology,* U. S. Government Printing Office, 1969.

30. W. L. Wolfe and G. J. Zissis, eds., *The Infrared Handbook,* U. S. Government Printing Office, 1984 (Available from SPIE).

31. J. Accetta and D. Shumaker, eds., *The Infrared and Electro-Optics Handbook,* ERIM and SPIE Press, 1993.

32. M. Bass, ed., D. Palmer, E. van Stryland and W. L. Wolfe, assoc. eds., *Handbook of Optical Principles and Practices,* McGraw-Hill, 1994.

33. Proceedings of the Institute of Radio Engineers, September 1959.

34. W. Smith and R. Fischer, eds., *Critical Review of Infrared Technology,* SPIE Press, 1993.

INDEX

 WILLIAM L. WOLFE was born in Yonkers, New York, at a very early age. He received his B.S. in physics from Bucknell University, cum laude. He did graduate work at the University of Michigan, where he received an M.S. in physics and an M.S.E. in electrical engineering. He held positions of Research Engineer and Lecturer at the University of Michigan. He was later Chief Engineer and Department Manager at the Radiation Center of Honeywell, in Lexington, Massachusetts. In 1969 he became Professor of Optical Sciences at the University of Arizona Optical Sciences Center, where he taught infrared techniques and radiometry. In 1995 he became Professor Emeritus. He has been a fellow and on the board of directors of the Optical Society of America; a senior member of the IEEE; and a fellow, life member, and past president of SPIE—The International Society for Optical Engineering. He is the coeditor of *The Infrared Handbook* and associate editor of the second edition of the *Handbook of Optics*. Present activities include ophthalmological instruments and optical techniques for cancer detection. He is the proud father of three wonderful children and as many grandchildren. In his spare time he fly fishes, sings, gardens, and uses his wife's telephone.